Value Solutions in Cooperative Games

Value Solutions in Cooperative Games

Roger A. McCain

Drexel University, USA

World Scientific

NEW JERSEY · LONDON · SINGAPORE · BEIJING · SHANGHAI · HONG KONG · TAIPEI · CHENNAI

Published by

World Scientific Publishing Co. Pte. Ltd.

5 Toh Tuck Link, Singapore 596224

USA office: 27 Warren Street, Suite 401-402, Hackensack, NJ 07601

UK office: 57 Shelton Street, Covent Garden, London WC2H 9HE

Library of Congress Cataloging-in-Publication Data
McCain, Roger A.
   Value solutions in cooperative games / by Roger A McCain.
   p. cm.
   ISBN-13: 978-9814417396 (hardcover : alk. paper)
   ISBN-10: 9814417394 (hardcover : alk. paper)
   1. Cooperative games (Mathematics) 2. Game theory. 3. Values. I. Title.
   QA272.4.M35 2013
   519.3--dc23

                             2012046042

British Library Cataloguing-in-Publication Data
A catalogue record for this book is available from the British Library

Typeset by Stallion Press
Email: enquiries@stallionpress.com

Printed in Singapore.

Foreword

This book is principally concerned with cooperative game theory. As such, it adopts several presuppositions. First, it is rational action theory. As such, it is a discourse about a logically possible world inhabited by "absolutely rational decision makers whose capabilities of reasoning and memorizing are unlimited" (Selten, 1975). It seems that this particular possible world is not instantiated precisely by the actual world. There is ample evidence that real human decision-makers are capable only of bounded rationality. There are two reasons to pursue rational-action theory nevertheless. First, rational-action theory provides the frame of reference without which bounded rationality cannot be defined. Second, theories based on bounded rationality may be subject to "theory absorption," as Morgenstern and Schwödiauer (1976) point out.

Further, cooperative game theory is a *particular kind* of rational action theory, invoking a distinct concept of rationality. Many (especially among economists) may hold that there is only one concept of rationality, the one common to noncooperative game theory and much of economics. That narrow view notwithstanding, (this book presupposes) cooperative game theory assumes rationality in a somewhat different sense: whenever agents can realize mutual benefits from choosing a joint strategy, they do so. *Rational* agents find ways to assure that their agreements are carried out and opportunism avoided. Rational agents also find ways to commit themselves to threats, even if, in the event, they will carry out the threats only with regret. But in the possible world of cooperative game theory, threats need never be carried out, since rational agents are aware that other rational agents will carry out their threats, so that agreements are always made that reflect the threats. The agents write complete contracts to specify their agreements.

In the possible world of noncooperative game theory (by contrast) rational agents always seize their best advantage in the momentary situations. Given opportunities they act opportunistically. Threats are never carried out if they are not "something [the threatener] would want to do, just for itself" (Nash, 1953). No agreements are carried out that are not self-enforcing. Agents write complete contracts to specify their agreements and the penalties for opportunistic violation.

Neither view is complete. In the actual world we observe opportunism and the seizure of momentary advantage, but we also observe agreements that reflect threats that are not "something [the threatener] would want to do, just for itself," such as the threat to strike. We also observe those threats being carried out, which would occur in neither of the possible worlds of game theory. Neither noncooperative nor cooperative game theory is likely to provide a good approximate description of this actual world, though *concepts from* both forms of theory seem necessary to construct such an approximate description. One segment of the literature on cooperative games that speaks to this complementarity is the literature on games in partition function form, since games represented in this way allow noncooperative interactions between coalitions, and games in partition function form will be a focus of this book as they were of Part II of McCain (2009).

Ideally, we would like to have a theory that would guide us as to when to apply cooperative, and when to apply noncooperative rationality. The neoclassical economics of the mid-twentieth century provided such a theory. It was assumed that interactions in "large groups" were always noncoop erative and that large group interactions determine important economic phenomena with few exceptions. For the exceptional small-group interactions, such as bilateral monopoly, cooperative concepts such as bargaining might be applied. (There was little consensus as to how the small group interactions might go, as illustrated by the muddle of oligopoly theory.) In the later twentieth century some such models were criticized, for example by Coase (1960) and Barro (1977) (to mention two that have been particularly influential) on the argument that rational agents would not, without good reason, forego potential gains from trade. Notice that this is precisely the cooperative concept of rationality. In response, cooperative elements (such as bargaining, again) have been brought ad-hoc into models that are thought of as noncooperative.

Nash (1953) had proposed the "Nash Program" that game-theoretic models of what appear to be cooperative phenomena should be noncooperative games with enforcement mechanisms, penalties and retaliations incorporated as strategic moves. While never made explicit, a Coase–Barro program would demand the opposite: that what appear to be noncooperative phenomena be modeled as cooperative games with explicit accounts of the barriers, such as transaction costs, that account for the failure to realize a fully cooperative solution. Unfortunately, neither Coase nor Barro seems to have had much knowledge of cooperative game theory (though Barro makes reference to "side payments") and the large literature of "microfoundations of macroeconomics" that has arisen from Barro's critique remains something of an incoherent muddle of cooperative and noncooperative rationality.

Nevertheless, this literature, particularly the discussion of labor market search and matching that was honored by the 2010 Nobel Memorial Prize, contains one important step toward a new, coherent combination of cooperative and noncooperative rationality. This step is the insight that, before they can coordinate their strategies as employers and employees, agents must invest their resources in establishing high-volume channels of communication with one another. And these resources unavoidably are committed noncooperatively, since people cannot cooperate until they are in communication.

Another contribution to such a coherent model is the Biform Game analysis due to Brandenburger and Stuart (1996, 2007). This is a two-stage model of value creation through coalition, with the first stage a noncooperative commitment and the second stage a cooperative one.

Both of these models share a difficulty with the core of a cooperative game, however, in that rational action may be consistent with a range of value imputations to the agents in the coalition or the firm. Brandenburg and Stuart explicitly use the core concept. In the search-and-matching models the wage may fall within a continuum of values.

This is an old problem in cooperative game theory, a problem that arose with the founding book with von Neumann and Morgenstern's (2004) solution set. Proposed solutions also have a long history. These may be denoted as "value solutions" following Shapley's (1953) contribution. (In fairness to economists, however) most of these solutions rely on simplifying assumptions in the representation of the cooperative game that distance cooperative game theory quite far from many applications. The main objective of this book, then, is to explore some extensions of some key value solutions that may lead toward a coherent theory of enterprise as it reflects cooperative rationality together with noncooperative commitments that occur before the links exist that would permit cooperative decision-making to occur. A very tentative model along these lines will be offered in the concluding chapters of the book. The central chapters focus on the determination of value solutions consistent with the core of a cooperative game and depending on the treatment of bargaining power.

Contents

Chapter 1

Value Solutions for Superadditive Transferable Utility Games in Coalition Function Form

Game theory has two major subdivisions: noncooperative and cooperative game theory. In noncooperative game theory, neither agreements to choose a joint strategy nor threats to retaliate or punish particular strategy choices are credible *per se*. That is, unless the threat or the terms of the agreement correspond to a subgame perfect Nash equilibrium, an agent who is rational in the sense of noncooperative game theory will not carry it out, and since this is commonly known, threats or agreement will have no impact on the actual decisions of the agents. In cooperative game theory, the assumption is instead that if an agreement is in the advantage of all those who are party to it, or if a threat can improve the payoffs to an agent, then rational beings will find ways to commit themselves to the agreement or the threat and enjoy the benefits that arise from the agreement or the credibility of the threat. The word "cooperative," used in this way, has no normative content.[1] Indeed, one possibility is that a group of agents may cooperate to exploit another agent or group of agents, as von Neumann stresses (1959, p. 33). This book will be concerned with cooperative games in this neutral sense.

There is a large literature on the topic. No attempt will be made to survey this literature as a whole. The purpose of this chapter is purely expository, and it will discuss some concepts that will be reconsidered and

---

[1]In Winter's words (2002, p. 11), "A cooperative solution concept is considered attractive if it makes sense as an arbitration scheme...some of the popular properties used to support solution concepts in this field are normative in nature." But this ambiguity can be avoided by treating the solution concepts as predictions about the formation of coalitions, and their sharing-out of mutual benefits, under particular definitions of rationality, where fairness concepts may or may not be considered as aspects of rationality.

extended in the balance of the book. Section 1.1 will summarize the logical framework for a very large and important segment of the literature, the theory of superadditive transferable utility (TU) games in coalition function form. Section 1.2 will discuss the concept of the core, which is not central to the purposes of this book but is the most widely used solution concept for these games and an important background for the study reported in this book. The remaining sections will discuss two concepts of solution that have a valuable property: for any such game, they provide a unique determination of the value that each agent can reasonably expect to derive from her or his participation in the game, i.e., *value solutions* for games in this category.

1.1.   Superadditive Games in Coalition Function Form

An agreement to choose a common strategy is conventionally called a *coalition* among the agents who are parties to the agreement. Any subset of the players in a game may constitute a coalition. A coalition of all players in the game is conventionally called *the grand coalition*. A coalition with only a single member is conventionally called *a singleton coalition*. Coalitions are usually considered to be non-null, but for some special purposes, it may be helpful to make reference to a *null coalition*, i.e., a coalition without members.

Noncooperative games may be represented in strategic normal or extensive (decision tree) form. Cooperative games are usually represented in a less detailed way, but the idea that the play of the game is the choice of a strategy, and in a cooperative game, a common strategy is always in the background. Accordingly, as a motivating example, consider the following game in strategic normal form[2]:

As we see, each player has three strategies. The unique Nash equilibrium occurs when the strategies played are (middle, center) and the payoffs are 6 and 4. At strategies (up, left), however, the payoff to Player b is nine times that at Nash equilibrium, while the payoff to Player a is only reduced by one-sixth relative to the Nash equilibrium. It seems that a shift from (middle, center) to (up, left) would make Player b better off by a big enough margin that he could compensate Player a for his sacrifice so that they both would be better off. The compensatory payment from b to a is called a *side*

---

[2]Some of these elementary points will be revisited in Section 8.1 in Chapter 8.

Table 1.1. Game 1.1. A game in strategic normal form.

		Player b		
First payoff to Player a, second to b		Left	Center	Right
Player a	up	5,36	3,3	3,3
	middle	6,3	6,4	4,3
	down	0,0	0,0	0,0

payment in game theory. This intuition is captured, in a strong and simple way, in a simplifying assumption widely used in cooperative game theory: transferable utility.

Assumption TU. In a coalition, the total payoffs to all players can be costlessly redistributed among the members of the coalition in any proportions.

Thus, having adopted Assumption TU, in case a coalition of Players a and b choose (up, left), the payoffs may be any pair that adds up to 41. Assumption TU rules out two kinds of complications. First, treating the payoffs as money payoffs, it may be that money is not equally a motivation to all players in all circumstances. The payoffs should be interpreted as utility payoffs, not as money payoffs, since otherwise in some circumstances rational players might not choose the largest payoff. But, second, there might be costs of making transfers, and particularly transfers in equal units of utility.[3] Assumption TU assures us that the payoffs are indeed proportionate to utility, but also that transfers can be costlessly made in ways that preserve the constant total utility accruing to the coalition. Accordingly, for the grand coalition of all players (in this case of the two players a and b) we can identify the *value* of the coalition as the largest total payoff that the two can obtain with a common strategy: in this case a value of 41, on the condition that they choose the joint strategy (up, left).

---

[3]This sort of informal discussion of the intended interpretation of formal assumptions is somewhat uncommon in the recent game theory literature, as Kaneko (2005) notes. He argues that this is unfortunate and leads to mutual misunderstanding exemplified by a Japanese comedy routine *Konnyaku Mondo*. A very rough American-English equivalent would seem to be "Who's On First?" (Abbot and Costello, 1938) In any case, von Neumann and Morgenstern, in contrast, were explicit in discussing the "plausible arguments" and interpretations of their formal constructions, and perhaps cooperative game theorists have avoided such discussions on the supposition that all that had been settled by von Neumann and Morgenstern. For the purposes of this book, however, some explicit comments on these matters will be needed, since the assumptions will be reconsidered in some chapters.

But what is the value of the singleton coalitions {Player a} and {Player b}? For each player, this will depend on the strategies chosen by the other player. It might seem that this would identify the values at the Nash equilibrium, and, as we will see later in the book, this could be a reasonable definition for certain purposes, but it is not customary in the literature on cooperative games. For, recall, "if a threat can improve the payoffs to an agent, then rational beings will find ways to commit themselves to...the threat." In Game 1.1, by choosing strategy (down), Player a can always reduce Player b to a payoff of zero. This will enhance Player a's bargaining power (we suppose), so he will make the threat and, if there is no cooperative agreement, will carry it out even at the sacrifice of a positive payoff for himself. Therefore, the greatest payoff that Player b can assure himself of is zero, and this is the value of the singleton coalition {Player b}. But, conversely, by choosing strategy (middle), Player a can assure himself of at least 4, and against (middle), the worst threat that Player b can make is strategy (right) which reduces Player a's payoff to 4 rather than 6. This threat makes Player b's bargaining power as great as it can be, and thus b will threaten (right) and the value of the singleton coalition {Player a} is the resulting value of 4.

We see that the values for the two singleton coalitions are based on inconsistent assumptions — that Player a chooses (down) for Player b's value but that Player a chooses (middle) for Player a's own value. But this, it seems, is the logic of rational threats.[4] Among cooperatively rational beings, no threat will be carried out anyway (since there is mutual benefit in coming to an agreement). Instead, both will see the logic of the situation and accede to a value consistent with the other player's most damaging threat. And this is the reasoning behind our second assumption:

Assumption AV. The value of a non-null coalition is the maximum (over its own strategies) of the minimum (over the strategies of all nonmembers)

---

[4]These conventions originate with von Neumann and Morganstern (2006), where they are supported by what the authors describe as plausible, not mathematical arguments. Together, they permit the generalization of the max min solution to general games, with appropriate reinterpretation. They were criticized as early as 1952 by McKinsey but have been rarely explicitly considered as against the alternative of Nash equilibrium or other concepts of noncooperative play among the coalitions. The discussion here has been influenced by Telser (1978). In Chapter 2, we reconsider the supposition that the agents always maximize their bargaining power by a threat that reduces the other coalition's value to the minimum possible.

Table 1.2.   Game 1.1 in coalition function form.

$\{a, b\}$	41
$\{a\}$	4
$\{b\}$	0

of total payoffs for the players in the coalition. That is, the value of the coalition is the assurance value (AV), the largest value the coalition can assure itself of against the most hostile of oppositions. (A null coalition always has a value of zero.)

Relying on Assumptions TU and AV, we may express Game 1.1 in terms of the values of all (three) possible coalitions, as in Table 1.2. Table 1.2 expresses game 1 in *coalition function form*. In general, for the theory of games in coalition function form, we need not be concerned with the choice of particular strategies and threats — these details can be left for operations research. Instead, for the balance of this chapter, we will be primarily concerned with games in coalition function form, defined as in Definition 1.1.

Definition 1.1. A game (in coalition function form) is a set $\{N, v\}$, where N is a set and $v : N^2 \to \Re$, i.e., v is a function from the set of all subsets of N to the real numbers.

Definition 1.2. For a *finite game*, the cardinality of N is a positive natural number.

Remark. This book will limit its attention to finite games, with exceptions as noted in Chapter 8.

A further assumption to characterize the family of games that is the focus of this chapter is *superadditivity*.

Definition 1.3. If Γ is a game in coalition function form and for any two coalitions, B and C, such that $B \cap C = \varnothing, v(B \cup C) \geq v(B) + v(C)$, then Γ is superadditive.

Here again, the assumption that cooperative games are superadditive originates with a "plausible" argument of von Neumann and Morgenstern and is rarely reconsidered. An exception is a paper of Aumann and Dreze (1974). It will also be argued that non-superadditive games may be interesting for pragmatic purposes, but that a (plausible) case can be made

Value Solutions in Cooperative Games

Table 1.3. Game 1.2. Another game in strategic normal form.

First payoff to Player a, second to b		Player b		
		Left	Center	Right
Player a	up	4,4	3,3	3,10
	middle	6,3	6,4	4,10
	down	10,0	10,0	10,10

Table 1.4. Game 1.2 in coalition function form.

$\{a, b\}$	20
$\{a\}$	10
$\{b\}$	10

that only superadditivity is consistent with cooperative rationality as it is understood in this chapter.

To motivate the discussion of superadditivity, consider Game 1.2, shown in strategic normal form in Table 1.3 and in coalition function form in Table 1.4. Here, strategies "right" and "down" assure each singleton of a value of 10 and thus define the values of the two singleton coalitions. These strategies also correspond to the unique Nash equilibrium for this game. Thus, the two singletons can attain the total payoff of 20 by independent action. If, nevertheless, they coalesce, they still have the option of choosing {down, right} as the joint strategy of the grand coalition, so indeed the value of the grand coalition can be no less than the sum of the values taken separately.

The example of Aumann and Dreze refers to farmers who, if they share their outputs, may as a result work "with less care and energy," so that the output of their joint work is less than the output would be if they worked separately. See McCain (2009, pp. 138–141) for a more detailed discussion. As McCain argues, Aumann and Dreze' interpretation is the more useful one if one's purpose is a pragmatic discussion of problems of collective farming.[5] However, in the context of cooperative rationality, we may observe that, if it is in the interest of both farmers to farm with more care and energy, they will be able to commit themselves to do so.

---

[5]Their discussion also resembles some discussion of the economics of incomplete contracts, a new topic in economics at about the time the Aumann and Dreze paper appeared, e.g., Williamson (1975) and Leibenstein (1969).

Nevertheless, there are some unstated assumptions in this reasoning. The unstated assumptions are, first, that any array of strategies available to the players acting independently is available also to a coalition of those players; and second, that there is no cost of forming a coalition and coordinating strategies *per se*. Neither of these assumptions is trivial. A relatively simple counterexample to the second is an example in which the players in the game have preferences with respect to the other players with whom they form coalitions. Suppose, for example, that Player *a* and Player *b* despise one another so much that neither will join a coalition with the other unless compensated by a utility increase of at least 1. They can still assure themselves of 10 by independently choosing {down, right}; so that determines the value of the singleton coalitions {Player *a*} and {Player *b*}; but the value of the grand coalition is 18.[6] Choosing coordinated strategies may also have resource costs in and of itself. Even if the strategies that would be chosen independently realize the greatest value for the coalition, this will have to be verified and communicated among them, and this may be costly. The example of Aumann and Dreze can be interpreted as follows: commitments to maintain an efficient effort within a coalition are particularly costly to attain and enforce, so that the farmers' coalition is actually better off without them, even though it cannot produce as much as the farmers can independently. Indeed, the assumption that coalition and the determination of a common strategy can be accomplished costlessly is not a trivial assumption.[7] Accordingly, for the purposes of this chapter and the next, we adopt a somewhat stronger assumption of cooperative rationality:

Assumption CCR[8] (Costless cooperative rationality). If an agreement is in the advantage of all those who are party to it, or if a threat can improve the payoffs to an agent or group of agents, then rational beings will find *costless* ways to commit themselves to the agreement or the threat and enjoy the benefits that arise from the agreement or the threat.

---

[6] A counter-critic might ask "would a *rational* being have such preferences?" The answer is that (transitive) preferences are incorrigible, so that for a general analysis, no preferences may be excluded.

[7] A parallel issue arises in welfare economics (see, e.g., Ng, 1980). A change of public policy that is a potential Pareto-improvement, i.e., that yields a positive balance of benefits over costs, may not lead to an actual Pareto-improvement unless losers can be compensated by lump-sum transfers. Such transfers may not, however, be possible. Here we are assuming, in effect, that lump-sum transfers are always possible within a coalition.

[8] In this chapter, we take Assumption CCR as the distinction between cooperative and noncooperative game theory. This will be further considered in Chapters 2 and 3.

The argument from Assumption CCR to superadditivity remains a plausible, not a mathematical argument. Of course, it could be formalized,[9] but that is not our objective here. The purpose is, rather, to sketch the boundaries of the universe of discourse of this chapter: it will not be applicable to cases in which a rational cooperative coordination of strategies can be costly.

This chapter (like the book) is primarily concerned with value solutions, but it will be helpful to sketch the concept of the core for these games. Although the core is not a value solution, it is probably the most widely applied solution concept for these games and captures some important intuitions about solution of cooperative games.

1.2. The Core

Let $\Gamma = \{N, v\}$ be a superadditive game in coalition function form. For such a game a *candidate solution* is an *imputation*, i.e., an N-vector of real numbers $y_1, y_2, \ldots$, such that $\sum_{i=1}^{N} y_i = v(N)$, with $y_i \geq v(\{i\})$; i.e., it is a set of imputed payments that exhaust the value of the grand coalition and that can be rationally accepted by each participant. Let $\mathbf{y}$ be a candidate solution and $B \subset N$.[10] If $v(B) > \sum_{i \in B} y_i$, then $\mathbf{y}$ *is dominated via B*. Let Y be the set of all undominated candidate solutions, i.e.,

$$Y = \left\{ \mathbf{y} \ni \sum_{i=1}^{N} y_i = v(N) \text{ and } \forall B \subset N, v(B) \leq \sum_{i \in B} y_i \right\}.$$

Then Y is the core of the game.

The core may be null, i.e., there may be no candidate solutions that are undominated.[11] Further, if the core is not null, it will often have a continuum of members. This poses the *core allocation problem* (Peleg and Sudhölter, 2003, p. 209): supposing that the core contains more than one imputation, which can a reasonable player expect? These well-known shortcomings of the core are inconvenient both mathematically and pragmatically, but nevertheless the core is valuable and important because it captures a key idea of competition in economics: if $\mathbf{y}$ is dominated via B,

---

[9]See, e.g., Peleg and Sudhölter (2003, p. 272).

[10]In this book $\subset$ will denote a proper subset, consistently with widespread convention, while a subset that may or may not be proper will be denoted by $\subseteq$.

[11]See, e.g., Forgo *et al.* (1999, p. 225).

we may express this by saying that group B have a competitive alternative that they find preferable to their share in **y**. The core comprises imputations for which no group has a preferable competitive alternative.

1.3. Shapley Value

In 1953, the *solution set* of von Neumann and Morgenstern was widely considered inadequate (McKinsey, 1952). Nash (1950) had published his bargaining theory, which was applicable only to two-person cooperative games. In that context, Shapley proposed his value solution for n-person TU games in coalition function form. [The core was proposed by Gillies (1953) in his doctoral dissertation at about the same time.] The Shapley value is a single imputation for any game in appropriate form. As such it was the first, and still the most influential, instance of a value solution for n-person cooperative games.

The Shapley value can be approached from two points of view: the computational and the axiomatic. From the computational point of view, consider a game $\{N, v\}$ with $|N| = n$. The value of each Player i in a game can be computed by the familiar formula,

$$\phi_i(v) = \sum_{S \subseteq N \setminus \{i\}} \frac{|S|!(n - |S| - 1)!}{n!} (v(S \cup \{i\}) - v(S)) \qquad (1.1)$$

where $\phi_i(v)$ is the value imputed to Player i. [In this book $\phi_i(v)$ will conventionally refer to the Shapley value of a game with coalition function v.]

The axiomatic derivation of the value establishes that it is the only imputation that satisfies certain conditions that may be thought of as reasonable. Two of these assumptions are that the solution is anonymous[12] and efficient (in senses that will be defined more precisely below). For Shapley's argument, however, the key assumption is additivity, which not only plays a central role in Shapley's proof, as we will see, but is also important for another reason. Shapley is interested in the simultaneous analysis of multiple games that may have overlapping memberships. Suppose, for example, that a, b, and c are considering a project to bring natural gas

---

[12]Shapley uses the term "symmetrical" but the term "anonymous" is more often used in later literature, and since Shapley also speaks of "symmetrical games," the modern term may make the discussion a little easier to follow.

home-heating fuel by pipeline to their homes on a remote lane, with sharing
of the overhead costs of the project; and meanwhile, c and d are considering a business partnership to develop new cell phone apps. Then, following
Shapley, the payoff to c should be the simple sum of his payoffs from the
natural gas project and the app development partnership. (Let us call this
case i.) Accordingly, Shapley characterizes the value imputation not over
the players in a single game, but over a universe U that is a superset of the
players in a particular game $\{N, v\}$. Any superset of N is a *carrier* of the
game $\{N, v\}$. The axiom system then is as follows.

Let the players in the universe be enumerated as $U = \{1, 2, \ldots u\}$ and
let π be a permutation of U. Then define the coalition function πv as
$\pi v(\pi S) = v(S)$.

$$\text{Axiom 1. Anonymity. } \phi_{\pi i}(\pi v) = \phi_i(v). \tag{1.2a}$$

For any carrier M of the game,

$$\text{Axiom 2. Efficiency. } \sum_{i \in M} \phi_{\pi i}(v) = v(M). \tag{1.2b}$$

Many subsequent discussions of the Shapley value have dispensed with the
discussion of carriers and assume efficiency only in the sense that the values
exhaust the coalition value of the grand coalition N. Shapley's assumption is equivalent, since the value of any agents not in N is zero, but
Shapley's approach assures us that his value assigns imputations consistently for games with overlapping memberships, so that the games can be
analyzed in isolation from one another, a powerful simplifying assumption.
Now let u and v be the coalition functions of two games. The sum of the
two games $(u + v)$ is the coalition function that assigns $u(C) + v(C)$ to
coalition C. Then we have

$$\text{Additivity. } \phi_i(u + v) = \phi_i(u) + \phi_i(v). \tag{1.2c}$$

Rather than reiterate Shapley's formal argument, we will sketch and illustrate it by a numerical example. The additivity axiom enables Shapley to
decompose any game in coalition function form into linear combination of
unanimity games. For a set of players N, the unanimity game is characterized by the value of 1 for any superset of N and zero for any set S with
$N \backslash S \neq \varnothing$. Shapley considers a generalization of the class of unanimity
games, which he calls symmetrical games, and first derives the value for
this class of games. A symmetrical game in this sense is a game $\{v, N\}$ that

Table 1.5. Game 1.3. A game
in coalition function form.

Coalition	Value
$\{a, b, c\}$	10
$\{a, b\}$	5
$\{a, c\}$	4
$\{b, c\}$	3
$\{a\}$	2
$\{b\}$	1
$\{c\}$	1

assigns a real number $c > 0$ to any superset of N. Anonimity and efficiency
are sufficient to assure us that the value for this game is $\frac{c}{|N|}$ for $i \in N$ and
zero otherwise.

Now consider the game in Table 1.5. We will express this game as a
linear combination of unanimity games or equivalently as a simple sum of
symmetrical games defined over the subsets of $\{a, b, c\}$.

Let v_R be the unanimity game over set R, i.e., v_R takes the value 1 for
any superset of R and zero otherwise. We posit

$$v = \sum_{R \subseteq N} c_R v_R \tag{1.3a}$$

where

$$c_R = \sum_{T \subseteq R} (-1)^{r-t} v(T) \tag{1.3b}$$

where $r = |R|$ and $t = |T|$. Thus, Eq. (1.3a) is a simple sum of a set of
symmetrical games. Consider, for example, $R = \{a, b\}$.

$$c_R = (-1)^{2-2} v\{a, b\} + (-1)^{2-1} v\{a\} + (-1)^{2-1} v\{b\}$$
$$= 5 - 2 - 1 = 2. \tag{1.3c}$$

Similarly the coefficients for $\{a, c\}$ and $\{b, c\}$ are both 1, and for the grand
coalition $\{a, b, c\}$ the coefficient is 2. For any singleton coalition $v\{j\}$ the
coefficient is the value $v\{j\ \}$. Individual a participates in the symmetrical
games defined over $\{a, b, c\}$, $\{a, b\}$, $\{a, c\}$ and $\{a\}$, with the respective
coefficients 2, 2, 1, 2. For the first of these coalitions, the value of Individual
a is $\frac{2}{3}$, for the second $\frac{2}{2}$, for the third $\frac{1}{2}$, and for the last one $\frac{2}{1}$. Thus, the
value of a in the original game must be the sum of these values for the

various symmetrical games, i.e., $4\frac{1}{6}$. Similarly we find that the values for b and c are $3\frac{1}{6}$ and $2\frac{4}{6}$.

Generalizing the reasoning in this example yields the permutational formula (1.1). Shapley's constructive proof is sufficient to establish the existence of the Shapley value for any game in appropriate form. (Shapley notes that the value can be computed for games that are not superadditive but may in that case not have the property of individual rationality: i.e., it is then possible that $\phi_i(v) < v(\{i\})$.) The additivity axiom also implies that it is unique. Certainly this is a remarkable result, and equally certainly, it depends strongly on the axiom of additivity.

The permutational formula (1.1) is subject to another interpretation. We note that it is a weighted sum of the values $(v(S \cup \{i\}) - v(S))$, i.e., the marginal contribution of Player i to coalition S. This property is often called *marginality*, and some alternative axiom systems for the Shapley value substitute marginality for additivity in the derivation of the value. (Winter, 2002, pp. 9–15) Shapley puts forward the following interpretation for the weights: let one of the permutations of the n players in the game be selected at random and the n players added to the coalition one at a time, with each receiving his marginal contribution. Then the weighted sum in Eq. (1.1) is the expected value of the individual's marginal contribution. This "bargaining model" for the Shapley value is of considerable interest in itself.

Marginality is thought of as a desirable property in that it corresponds to the economic concepts of marginal productivity and utility. In economics, however, payment according to marginal productivity is a property of a general competitive equilibrium, a model that presupposes competition among many coalitions called *firms* for a limited number of homogenous resources and for customers to buy a limited number of homogenous products. If the grand coalition were to form by merging all these firms, and enlisting their resource suppliers and customers as members, then marginal productivity payments are no longer predictable. But this grand coalition is what the theory of superadditive games in coalition function form presupposes.

Additivity (or marginality) is a defining property of the Shapley value. Is, then, additivity a reasonable property to require of a solution? In the example considered above, a, b, and c are considering joint action to bring natural gas to their homes as a heating fuel, while c and d are exploring a business partnership to generate some income by selling cell phone apps. In such a case, it is reasonable to suppose that the payoff to c should be the sum of the benefits to c from the two different interactions. Suppose, instead, (case ii) that c and d are negotiating for a geothermal heating system to be installed in c's home by d, a contractor. In this case it hardly makes sense to suppose that the payoffs to c are the sum of the payoffs in the two different games. The installation of the geothermal heat system would eliminate one major use c would otherwise have for natural gas, reducing the value of the gas pipeline for c, and conversely; so the sum of his payoffs in both games must be more than the payoff in the game that comprises both interactions. Yet again, (case iii) suppose that c and d are negotiating terms of a contract for d to replace c's existing electrical heating system with a gas-fired system. The value to c of this coalition will be zero unless the other coalition takes place, but may be quite substantial if the coalition of $\{a, b, c\}$ is successful. Again, it would not be reasonable to suppose that the values of these two games are additive. We should stress that this does not suggest any inconsistency in Shapley's logic: rather it tells us that Shapley's logic does not apply to a "universe" that includes games like cases ii and iii.

Put otherwise, the additivity axiom excludes any cases in which outcomes of some coalitions are either substitutable or complementary. Conversely, the Shapley value is applicable in a world in which there are no complementarities or substitutabilities in the objectives of coalitions in different games. In a world in which substitutabilities and complementarities are common, games cannot be decomposed into smaller games as Shapley wishes to do. It has been known for generations[13] that complementarity can create a problem with marginal imputation when complementarity is sufficiently strong to result in increasing returns to scale. In that case, the marginal imputations total more than the value of the aggregate. Further, strict superadditivity corresponds to increasing returns to scale. Thus, problems are to be expected in the presence of complementarity.

Finally, should we consider the Axiomata (1.2a)–(1.2c) as normative or otherwise? Axiom (1.2.a). seems to follow directly from cooperative rationality. However, Axioms (1.2.a) and (1.2.c) seem best regarded as characterizing the rationality, in some sense, of the solution itself. Axiom (1.2.c), in particular, has been described as an assumption of "accounting rationality" of the solution (Forgo *et al.*, 1999, p. 258.) It seems an open question to what extent the rationality of the solution as a whole can be induced from the cooperative rationality of the agents.

---

[13]See von Wieser (1889).

Regardless of these critical points, the Shapley value is a remarkable achievement and quite properly a model for much subsequent research in game theory.

1.4. The Nucleolus

Over the 15 years following 1953, a number of other concepts for solutions of cooperative games were proposed. In 1969, Schmeidler offered a solution he called the *nucleolus* of the game. A key step in the computation of the nucleolus was the definition of the *excess function*, which we will take up next. A few years later, Aumann and Dreze (1974) showed that the excess function could be used to characterize a number of solution concepts previously discussed, including the core but not the Shapley value.

Let $\Gamma = \{N, v\}$ be a superadditive game in coalition function form, $\mathbf{x}$ an imputation, and $S \subset N$. Let $\mathbf{x}_S = \sum_{i \in S} x_i$. Then the excess function is

$$E(\mathbf{x}, S) = v(S) - \mathbf{x}_S \qquad (1.4)$$

The excess for S can be thought of as a measure of the discontent of the members of S with the imputation $\mathbf{x}$. If the excess is nonpositive for every set S, then $\mathbf{x}$ is an element of the core of Γ.

Since we have computed the Shapley value for Game 1.3, we may use it to illustrate the concept of the excess. The Shapley value imputation for Game 1.3 is $\mathbf{x} = 4\frac{1}{6}, 3\frac{1}{6}, 2\frac{4}{6}$. Thus, $E(\{a, b\}, \mathbf{x}) = 5 - (4\frac{1}{6}) - (3\frac{1}{6}) = -2\frac{1}{3}$. By similar calculations we obtain the excess function for this game and imputation as shown in Table 1.6.

Table 1.6. The excess function for the Shapley value for Game 1.3.

Coalition	Excess
$\{a, b\}$	$-2\frac{1}{3}$
$\{a, c\}$	$-2\frac{5}{6}$
$\{b, c\}$	$-2\frac{5}{6}$
$\{a\}$	$-2\frac{1}{6}$
$\{b\}$	$-2\frac{1}{6}$
$\{c\}$	$-1\frac{4}{6}$

**Table 1.7. The excess
function for Game 1.3
with imputation 4,3,3.**

Coalition	Excess
$\{a, b\}$	-2
$\{a, c\}$	-3
$\{b, c\}$	-3
$\{a\}$	-2
$\{b\}$	-2
$\{c\}$	-2

Schmeidler then defines the nucleolus as the imputation $\mathbf{y} = \psi(v)$ that satisfies

$$\min_{\mathbf{x}} \max_{S} E(S, \mathbf{x}). \qquad (1.5)$$

We can immediately verify that for Game 1.3, the Shapley value differs from the nucleolus. For the Shapley value, the singleton $\{c\}$ is the coalition with the largest excess. This excess can be reduced by transfers to c from a and b. Suppose in particular that a and b each transfer $\frac{1}{6}$ to c, yielding the imputation 4,3,3. The excess function is then as shown in Table 1.7.

We can readily verify that this is the nucleolus for this (very simple) game. Any shift away from the imputation 4,3,3 will make one of a, b, and c worse off, increasing the excess above -2 for the singleton coalition that will then have the largest excess.

In this book $\psi_i(v)$ will conventionally refer to the nucleolus of a game with coalition function v. It is sometimes useful to refer instead to the pre-nucleolus, an extension of the nucleolus that includes values $\psi_i(v) < v(\{i\})$, i.e., values that may not be "individually rational."

Intuitively, we may think of the nucleolus in the following terms: consider a bargaining process in which a feasible demand by a group that is "more discontented" than the residual group will be followed by a concession from the residual group. Rational self-interest will lead the most discontented group to demand an adjustment in their favor, and so long as they remain the most discontented group, this will continue. Such a process would stabilize at the nucleolus.

It is of some interest that the nucleolus is more nearly equal than the Shapley value in this case. We should also note that the nucleolus, as defined here and in Schmeidler's paper, can be thought of as one instance of a larger family of value solutions based on the excess. In expression (1.5), the vector of excesses of potential deviating coalitions is aggregated by the max operator, and the aggregrate value is minimized. More generally, as suggested by Forgo *et al.* (1999, p. 252), we might choose a different aggregation formula, such as the variance of the excesses. This is suggested on the basis of a normative judgment in favor of still greater equality.

As with the Shapley value, there are several axiomatizations of the nucleolus, and one will be of particular interest for comparative purposes.

1.5.   A Comparative Perspective

An axiomatic derivation of the pre-nucleolus would begin with anonymity, i.e., axiom 1, expression (3a). A second axiom for the pre-nucleolus is covariance under strategic equivalence. Let $\Gamma = \{N, v\}$ and $H = \{N, w\}$ be games and for every $S \subseteq N$, **a** a vector of n constants and b a real constant, and $w(S) = bv(S) + \sum_{i \in S} a_i$, then Γ and H are said to be *strategically equivalent*. Let $\varphi(v)$ be a value solution for games in coalition function form. Then

Axiom 4. Covariance.   If Γ and H are as indicated above,

$$\text{then } \varphi_i(w) = b\varphi_i(v) + a_i.$$

Essentially, Axiom 4 says that the units of measurement for the values of coalitions do not matter. If we substitute a measurement with different units and a different zero point, we simply transform the solution by the same constants, and obtain the solution for the transformed game. Alternatively, Axiom 4 identifies the payoffs in the game as utilities in the von Neumann–Morgenstern sense. Note that the Shapley value, like the nucleolus, has this property.

For the third axiom for the prenucleolus, we need a concept of a *reduced game*. In general, suppose some commitments have been made with respect to payments to be distributed to a subset of the players in the game. The subset whose payments are settled are $N \backslash S$. but it remains to settle the payments of S. Payments to $N \backslash S$ are consistent with a feasible imputation **y**. We then define the reduced game v_S for subsets $T \subseteq S$ by

$$v_S(T) = \begin{cases} 0 & \text{if } T = \varnothing \\ v(N) - y(N \backslash T) & \text{if } T = S \\ \max_Q \left(v(T \cup Q) - \sum_{i \in Q} y_i : Q \subseteq N \backslash S \right) & \text{otherwise.} \end{cases} \tag{1.6}$$

The reduced game is $\{S, v_S\}$. Suppose then that $\varphi(v)$ is a value solution on $\Gamma = \{N, v\}$ and $\mathbf{y} = \varphi(v)$. If then $\varphi_i(v_S) = \varphi_i(v)$, for $i \in S$, then $\varphi(v)$ is said to have the reduced game property or the property of consistency over the reduced game. This is Axiom 5:

Axiom 5. Reduced Game Property (1). $\varphi(v)$ has the reduced game

property with the reduced game defined as in (1.6).   (1.7)

From Axioms 1, 4, and 5, it may be shown that $\varphi(v)$ is the pre-nucleolus (Winter, 2002, p. 13). We recall that the pre-nucleolus may include values that are not individually rational. Accordingly, we may modify 7 to limit the solution to values that are individually rational, i.e., to imputations (Forgo *et al.*, 1999, p. 252). A similar axiomatic derivation then yields the nucleolus.

Suppose instead that we define the reduced game as follows. Once again, payments are settled for $N\backslash S$, the settled payments to $i \in N\backslash S$ are denoted by y_i and the reduced game is the game $\{w, S\}$, such that

$$w_S(T) = v(T \cup (N\backslash S)) - \sum_{i \in N\backslash S} y_i. \qquad (1.8)$$

We define consistency as before.

Axiom 6. Reduced Game Property (2). $\varphi(v)$ has the reduced game

property with the reduced game defined as in (1.8).   (1.9)

From Axioms 1, 4, and 6, it may be shown that $\varphi(v)$ is the Shapley value (Winter, p. 12).

We see that the nucleolus and the Shapley value share Axioms 1 and 4, i.e., anonymity and covariance over strategically equivalent games; and share a property of consistency over reduced games, with the qualification that reduced games are differently defined. The reduced game is played over subset S. In (1.8), each coalition within S is valued on the basis that it is collaborating with the entirety of the group for whom the payments are settled and is able to distribute among those who are members of S whatever coalition value is left from the already settled distribution to the members of N/S. In (1.6), the assumption is instead that a coalition within S is valued on the basis of collaboration with some subset of $N\backslash S$, and the subset chosen is the one that the coalition within S can most profitably collaborate.

The reduced game property, in either case, is an assumption on the rationality of the solution. Yet both definitions of the reduced game seem somewhat arbitrary. Equation (1.8) is suggestive of double-counting. Let $T \subset S, U \subset S, T \cap U = \varnothing$. Then both T and U are valued on the assumption that each can appropriate any net value from $N \backslash S$. But both cannot do it. Equation (1.6) may or may not imply double counting of this kind and may further be inconsistent in that different coalitions within S may be valued on the basis of different subsets of $N \backslash S$, which may or may not overlap. In the case of (1.6), we may envision a coalition T calling a committee meeting to determine what is the largest value it might demand, somewhat in the spirit of the excess measure. In the case of (1.8), we might suppose that the committee estimates its entitlement by the value it is able to add to the settled group $N \backslash S$, in the spirit of the marginality condition for the Shapley value. In any case, the assumption of consistency over the reduced game is an assumption, in effect, that the value solution itself cancels this possible irrationality.

1.6. Chapter Summary

A very large proportion of the literature on cooperative games is focused on superadditive TU games in coalition function form. These games are defined by powerful simplifying assumptions, and while these assumptions can be supported by persuasive plausible arguments, the assumptions serve to define the boundaries of the domain of applicability of the literature.

Among these assumptions, the transferability of utility, Assumption TU, is pragmatically useful if we wish to focus on the side payments by which payoffs may need to be redistributed in order to assure cooperation. Together with Assumption AV, which is that a coalition will always face the threat of the most hostile of possible opposition strategies, this enables us to identify the unique value of each coalition, so that any game can be expressed in coalition value terms. The cooperative solutions can be thought of as rational in the sense of costless cooperative rationality, Assumption CCR.

For this class of games, the core is the most common solution concept, but as the core may be null or may contain many candidate imputations, it is often indecisive. By contrast, value solutions are solutions that yield a single imputation as the solution. Of these, the Shapley value is the best known. It relies among others on an axiom of additivity which seems to exclude any universe of games in which some of the games to be added seek objectives that are complementary or substitutable. Somewhat less known is the nucleolus (and the pre-nucleolus). Both the Shapley value and the pre-nucleolus can be derived from axiomata of anonymity, covariance over strategically equivalent games, and consistency of valuation of reduced games, where the reduced games are differently defined. However, both definitions of reduced games have some element of arbitrariness. In general, as in this case, it seems that the choice among cooperative game solution concepts is a choice among alternative concepts of the rationality of the solution, none of which is clearly necessary for individual rationality even in the very narrow sense we have adopted in this chapter.

Chapter 2

Zeuthen–Nash Bargaining

While bargaining theory is sometimes treated as a subfield of game theory, it is more accurately described as an independent field that overlaps game theory. The earliest work (Zeuthen, 1930; Hicks, 1932; Pen, 1952) preceded the widespread awareness of game theory, and important work continued outside game theory (e.g., Bishop, 1964; Coddington, 1966; Cross, 1975; Saraydar, 1965). Union-management bargaining over a wage rate was often treated as the paradigmatic case. However, the contributions of John Nash (1950, 1953) to bargaining theory as game theory (or conversely) have been as widely influential as any, and give the earliest example of a value solution for a class of cooperative games, together with a call that value solutions should be applicable to all cooperative games (Nash, 1950, p. 157). Thus Nash's theory, and some of its generalizations, will be central to the purposes of this book.

2.1. Nontransferable Utility Games

Nash (1950, p. 155) addresses a situation in which "two individuals...have the opportunity to collaborate for mutual benefits in more than one way." The different forms their collaboration may take result in different assignments to the two individuals of the benefits from collaboration. Either there are no side payments or it may be impossible by means of side payments to reallocate utility among the agents unit for unit. Thus, Assumption TU is not applicable, and Nash's bargaining theory is applicable to *two-person nontransferable utility* (NTU) games. In fact, TU games are a special case of NTU games, and Nash bargaining can be applied to two-person TU games; however, the extension to NTU games is important.

As before, a numerical example will illustrate this. Consider Game 2.1, shown in strategic normal form in Table 2.1. Clearly, {middle, center} is the

Table 2.1.   Game 2.1.

First payoff to Player a, second to b		Player b		
		Left	Center	Right
Player a	up	10, 10	2, 11	3, 3
	middle	21, 2	7, 5	0, 3
	down	2, 4	2, 0	2, 3

unique Nash equilibrium for this game, and Player a can assure himself of
2, and Player b of 3, by playing down and right, respectively. Players a and
b can obtain a Pareto improvement[1] over either of these strategy vectors
by choosing {up, left}. In addition, the players can play many mixed and
correlated[2] strategies, and in particular, some correlated strategies that
assign positive probabilities to {up, left} and {middle, left} and some that
assign positive probabilities to {up, left} and {up, center} will also result
in Pareto-improvements. Nash assumes von Neumann–Morgenstern utilities
and risk neutrality, (1950, p. 157, expression b), so the expected values are
certainty equivalent in utility terms.

We can visualize the different ways the two players can collaborate by a
Cartesian graph, with the utilities or mathematical expectations of utilities
on the two axes. Consider Fig. 2.1.[3] The utility or payoffs to Player a are on
the horizontal axis, and those to Player b on the vertical axis. Points α, β,
and γ show the pure strategy vectors {up, center}, {up, left} and {middle,
left}, respectively. Points along the line $\alpha\beta$ show correlated strategies that
assign positive probabilities to strategy vectors {up, center} and {up, left},
while the points along $\beta\gamma$ show correlated strategies that assign positive
probabilities to strategy vectors {up, left} and {middle, left}. These lines
form the upper boundary of the region of expected utility vectors that the
two agents can attain. The region is bounded below and to the left because

---

[1] To say that a shift of strategies creates a Pareto improvement is to say that it makes at
least one player better off and the other or others no worse off; that is, in the terms used
in the previous chapter, the situation after the shift Pareto-dominates the one before.

[2] A correlated strategy is a strategy vector that is jointly mixed, i.e., in a two-person
game such as this one, a strategy that assigns positive probabilities to two or more
strategy vectors. Much of the literature on correlated strategies has been in the context
of noncooperative games and so is particularly concerned with correlated strategies that
are self-enforcing; but here we are assuming that the agents can commit themselves to
any feasible strategies, so that the noncooperative limitations on correlated strategies do
not apply.

[3] This discussion is influenced by Luce and Raiffa (1957).

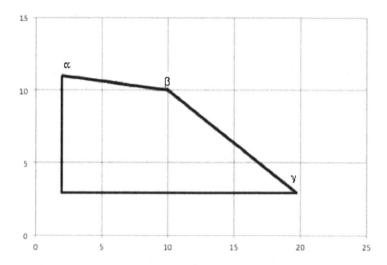

Fig. 2.1.   Utility possibilities in Game 2.1.

the two players cannot be reduced below utilities of 2 and 3, respectively. In any case, players who are cooperatively rational will consider only points along the upper boundary of the set.

The region bounded by $\alpha\beta$ and $\beta\gamma$ would be referred to in welfare economics (e.g., Ng, 1980,) as a utility-possibility frontier. As shown, the feasible region is compact and weakly convex. In a more complex game, a utility-possibility frontier may be generated by non-probablistic arrangements. For example, in a cartel, some members may be allowed to price below the joint profit-maximizing price schedule, *in lieu* of a side payment, to draw them into the cartel. In such a case, the frontier could be smooth and defined in deterministic terms.

This provides one approach to an analysis of games without the assumption of transferable utility, i.e., NTU games. For TU games, a coalition can be characterized by the total value it creates, and this is the beginning point for a cooperative-game analysis. For NTU games, the utility-possibility frontier for each particular coalition provides the corresponding starting point.

(Yet another approach, where side payments are logically excluded, will simply enumerate outcomes that a particular coalition can bring about: this enumeration is the effectiveness set for the coalition. Effectiveness sets will play no role in this book.)

Convexity may not be assured in the absence of diminishing marginal returns and risk neutrality, but this case is far less tractable and conditions sufficient for convexity are usually assumed in the economic applications. Thus, in much of the literature on nontransferable utility, coalitions of two or more players are characterized not by well-defined values but by compact, convex utility regions. And this is Nash's assumption.

Each point on the upper boundary is Pareto optimal,[4] but there are many Pareto optima — in general, as in this case, the set of Pareto optima will constitute a continuum in the space of utility imputations. This had been known since the 1890s. We suppose that rational bargainers will come to an agreement that is Pareto optimal. But which? This is the *bargaining problem*.

2.2. Zeuthen's Solution

Zeuthen (1930) had approached the bargaining problem from the point of view of rational assessment of risk and expected values. Suppose that two bargainers are to decide the values of u and v, where u is the utility of the first bargainer, v of the second, and u and v are constrained by $f(u,v) < 0$. In the absence of an agreement the first bargainer can be assured of d_u and the second of d_v. The function $f(u,v)$ is assumed to have properties sufficient that the utility possibility set is convex. Suppose that bargainer 1 faces an offer of u_1 and is considering adopting a minimum demand of $u_2 > u_1$; but that if he does so, the risk of a breakdown of bargaining is p. Then he will adopt the holdout strategy only if

$$pd_u + (1-p)u_2 \geq u_1. \tag{2.1}$$

Conversely, the greatest risk that bargainer 1 will accept is

$$p = \frac{u_2 - u_1}{u_2 - d_u}. \tag{2.2}$$

Similarly, the greatest risk bargainer 2 will accept is

$$q = \frac{v_2 - v_1}{v_2 - d_v}. \tag{2.3}$$

Zeuthen argues that each bargainer will concede, reducing u_2 or v_2, whenever the other bargainer will take greater risk. Thus a condition for stable

---

[4]That is, undominated. More explicitly, a situation is Pareto-optimal if there is no shift of strategies or arrangements that can generate a Pareto-improvement from it.

minimum demands is $p = q$, that is,

$$\frac{u_2 - u_1}{u_2 - d_u} = \frac{v_2 - v_1}{v_2 - d_v} \tag{2.4a}$$

equivalently,

$$\frac{u_2 - d_u}{v_2 - d_v} = \frac{u_2 - u_1}{v_2 - v_1} \tag{2.4b}$$

as the bargaining proceeds, $u_1 \to u_2$ and $v_1 \to v_2$, so that, in the limit,

$$\frac{u_2 - d_u}{v_2 - d_v} = \frac{\partial f / \partial v}{\partial f / \partial v} \tag{2.4c}$$

and this is fulfilled if u_2 and v_2 are such as to

$$\max(u_2 - d_u)(v_2 - d_v) \quad \text{subject to } f(u_2, v_2) \le 0. \tag{2.5}$$

Zeuthen makes a point that he assumes no differences in bargaining power. However, Zeuthen assumes that the utilities are measureable as "cardinal utilities." He does not assume risk neutrality, and the alternatives described by the condition $f(u, v)$ are deterministic.

2.3. Nash Bargaining with Equal Bargaining Power

As we have seen, Nash (1950) assumes von Neumann–Morgenstern utilities and risk neutrality (*"linearity"* of the utility functions) and assumes that the solution should, accordingly, be independent of linear transformations of the utility function. He makes three further assumptions about *the rationality of the solution.* They are (1) Pareto optimality, (2) independence of irrelevant alternatives, and (3) a symmetry assumption. While the symmetry assumption is not made quite explicit, it is clear that what Nash means is that if (u, v) is at the boundary of the utility possibility set, so is (v, u). Nash's assumption is that if the game has this symmetry property, the bargain should award equal payoffs to both parties, and he describes this as an assumption of equal bargaining power. Pareto optimality is justified as arising from the individual rationality of the bargainers. We will return below to the axiom of the independence of irrelevant alternatives.

Nash then states the theorem that the unique solution consistent with these axiomata is characterized by (2.5). His proof is informally sketched and is as follows: first, the convexity and compactness of the utility possibility set assures that such a maximum exists and is Pareto optimal. Second, transform the utility functions so that $d_u = d_v = 0$ and $u_2 = v_2 = 1$. Nash then considers an alternative bargaining game in which the utility possibility frontier is a supporting hyperplane tangent to $f(u_2, v_2)$ at the maximum point. This larger alternative game is symmetrical, so that its solution must give equal payoffs to the two bargainers, and the only equal payoffs in the Pareto optimal bounding set for the alternative game are 1,1. This must be the solution of the bounding game, and so, by the independence of irrelevant alternatives, must also be the solution for the original game. For Game 2.1, Nash's solution is shown by Fig. 2.2. The hyperbola shows the locus of payoffs at which $(u_2 - d_u)(v_2 - d_0) = 56.48011$. At the solution, the two bargainers play a correlated strategy solution assigning probability 0.9261 to {up, left} and 0.0739 to {middle, left}.[5] The expected

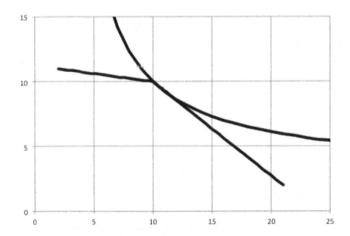

Fig. 2.2.   Nash's solution to Game 2.1.

---

[5]This should be taken strictly as an illustrative example. Nash remarks "the 'fair bargain' might consist of an agreement to use a probability method to decide who is to gain in the end." Applying this to Game 2.1, the bargainers might choose {middle, left}, so that a collects 23, and further agree to a wager in which a would transfer the 23 to b with a probability between zero and $1 - \frac{2}{21}$. The resulting utility possibility frontier dominates the one shown in the figure. Formally, Game 2.1 plus a wager is a different game, so there is no strict contradiction here, but Nash gives us little guidance as to how indeed we might understand an NTU game if individuals are rational and utilities are indeed linear as he assumes.

value payoffs are then 10.8129 for the first bargainer and 9.4088 for the second.

Clearly, the independence of irrelevant alternatives is crucial for Nash's argument. Nash gives a plausible argument for it, as follows: Assuming $S \subset T$, "If two rational individuals would agree that $c(T)$ would be a fair bargain if T were the set of possible bargains, then they should be willing to make an agreement ... not to attempt to arrive at any bargains represented outside the set S if S contained $c(T)$." To "attempt to arrive at any bargains outside of S" is to threaten to refuse the bargain unless a payoff pair outside S is offered (Nash, 1950, p. 159). Thus we see that the independence of irrelevant alternatives, in the context of bargaining theory, is an assumption that a "rational" person will not use the threat possibilities in an unsymmetrical utility possibility set to enhance his bargaining power.

To illustrate the point, recall that the Nash bargaining theory can be applied to transferable utility games as a special case, and recall Game 1.1 in coalition function form, as shown in Table 1.2 in Chapter 1. The Nash solution is equal division at 20.5 for each bargainer [and indeed equal division is the Nash solution for all TU games for which there is an interior solution to the maximum at Eq. (2.5)]. We may contrast this to the Shapley value and the nucleolus for Game 1.1. For this game, the Shapley value and the nucleolus are identical with imputations of $22\frac{1}{2}$ for Player a and $18\frac{1}{2}$ for Player b. This reflects the fact that a can assure himself of 4, so that his marginal contribution and his excess both are greater than those of b in given circumstances. Recall that a can threaten to reduce b to zero, while b can at worst threaten to reduce a to 4. Now suppose we substitute for Game 1.1 a larger game in coalition function form in which the value of each of the singleton coalitions is zero. The utility possibility frontier for this game will be a superset of that for Game 1.1 and will include the Nash solution point for Game 1.1, but no point Pareto-preferable to it; so, by the independence of irrelevant alternatives, must have the same solution. For the Shapley value and the nucleolus, however, the solution will differ from that for Game 1.1. Since the larger game is symmetrical, the marginal contributions and excesses are symmetrical, and so both the Shapley value and the nucleolus will impute equal payoffs of 20.5 to the two players. This corresponds to the fact that in the larger game, the threat possibilities are symmetrical. We see that the independence of irrelevant alternatives conflicts with both marginality and any influence of unsymmetrical threat possibilities away from the solution point.

We see, then, that independence of irrelevant alternatives cannot indisputably be deduced from individual rationality. Indeed an argument can be made that it conflicts with individual rationality. We should note that Nash does not assume cooperative rationality as defined in Chapter 1. In Nash (1953), he explicitly considers the alternative threat behaviors of the bargainers, proposing a theory of variable-threat bargaining. He says, "Supposing a and b to be rational beings, it is essential for the success of the threat that a be *compelled* to carry out this threat if b fails to comply. Otherwise it will have little meaning. For, in general, to execute the threat will not be something a would want to do, just for itself... we must assume there is an adequate mechanism for forcing the players to stick to their threats and demands once made, and one to enforce the bargain, once agreed" (Nash, 1953, p. 130; emphasis in the original). In other words, we require not only enforceable agreements but also enforceable threats. The concept of rationality Nash sketches here is noncooperative rationality and is the predominant hypothesis of rationality in noncooperative game theory and modern economics. On the face of it, the idea of enforceable threats is less clear than that of enforceable agreements: it is difficult to imagine a party to a negotiation being sued because he did not carry out a threat. In some cases, though, it may be possible to incorporate threats in a contract, as when a performance bond is required. In any case, if *all conceivable* threats and contracts can be *costlessly* enforced, then noncooperative and costless cooperative rationality would be equivalent, and that is clearly Nash's intention. In the axiomatic proof of this more general bargaining theory, the independence of irrelevant alternatives again serves the purpose of excluding threats and demands at distance from the solution point. But it remains unclear why any person who is rational — either in Nash's noncooperative sense or in the cooperative sense — would foreswear such threats if he can benefit from them by increasing his bargaining power.

Cooperative rationality is an extreme assumption about human behavior: so is noncooperative rationality. This is most evident in the case of threats. In a spirit of casual empiricism, it does seem that history offers many examples of people carrying out threats that were not "something [the threatener] would want to do, just for itself." People are sometimes spiteful. Perhaps if we look carefully we will also see examples of agreements carried out in the absence of compulsion. But we do also observe bluffs and cheats. Neither extreme assumption is always a good approximation to real human behavior. Similarly the assumption that *all* agreements and threats can be *costlessly* enforced is an extreme assumption, as is the assumption in Nash's noncooperative game theory that "it is impossible for the players to communicate or collaborate in any way" (Nash, 1953, p. 129), as Nash indeed observes.

Nash's solution has some very appealing properties. However, the crucial property that defines it, the independence of irrelevant alternatives, is known to be quite problematic, and it is problematic in particular in that it imposes limits on the threat behavior of the bargainers, assuming that asymmetrical threats away from the final solution point either cannot or will not be used to shift the solution point.

2.4. Variable-Threat Bargaining

The bargaining theory derived in Nash (1953) has been called a *variable-threat* bargaining theory. It is a two-stage decision model. In the first stage, the bargainers choose their *threats* from a defined set of available threats, which may be mixed strategies. These threats then determine the disagreement point. At the second stage, then, the bargaining is decided by the bargaining theory of Nash (1950), relative to the threat points determined by the first stage. Implicitly applying backward induction, the threats chosen at the first stage then define a noncooperative game on the threat strategies. The solution to the variable-threat bargaining game is a Nash equilibrium in this first-stage reduced game. Nash's Axiomata VI and VII serve to establish that the solution to the two-stage game is the same as the solution to a game in which the bargainers can choose only the Nash equilibrial threats, and the theory of Nash (1950) is directly applicable to that restricted game. Thus "the solution of the game not only gives the utility of the situation to each player, but also tells the players what threats they should use in negotiating."

This is an important advance: von Neumann and Morgenstern had *assumed* that the most hostile threat that an individual or coalition would make would give the best advantage in bargaining. But this may not be true. Consider Game 2.2, in which the players first play the coordination game shown in Table 2.2 and then redistribute the payoffs via costless side payments. For Game 2.2 the bargaining utility possibility frontier is determined by

$$u + v - 20 \leq 0. \tag{2.6}$$

Table 2.2. Game 2.2.

First payoff to a, second to b		b	
		S_1	S_2
a	S_1	10, 10	4, 0
	S_2	0, 4	5, 5

Table 2.3. Threat outcomes for Game 2.2.

Threat outcomes for a, then b		b	
		S_2	Mixed
a	S_2	5, 5	4.545, 0.909
	mixed	0.909, 4.545	0.826, 0.826

Table 2.4. Bargaining solutions for Game 2.2.

First payoff to a, second to b		b	
		S_2	Mixed
a	S_2	10, 10	12.73, 7.27
	mixed	7.27, 12.73	10, 10

For Game 2.2, the most hostile *pure* strategy is S_2: if a adopts S_2 then b can assure himself of no more than 5, while for S_1 b can assure himself of 10. However, by adopting a mixed strategy that assigns a probability of $\frac{1}{11} = 0.090909$ to strategy S_1, a can prevent b from assuring an expected value payoff more than 4.545, and this is the most hostile threat that agent a can make from the set Nash considers as eligible. It is the strategy von Neumann and Morgenstern assume that rational players would choose. Table 2.3 compares the threat outcomes for pure strategy S_2 with those for this mixed strategy. Table 2.4 shows the bargaining outcomes for S_2 with the mixed strategy. Table 2.4, in other words, shows the reduced game with bargaining payoffs inserted that correspond to the second stage bargaining outcome of the game, but considering only the selected strategies S_2 and the von Neumann–Morgenstern mixed strategy. We see that the choice of the most hostile threat does not enhance a's bargaining power, since in fact this threat is a dominated strategy in the bargaining power game. This occurs because the most hostile threat is the one that reduces the disagreement payoff of the threatener even more than it does that of the victim of the threat. On this point, Nash's assumptions seem to agree better with individual rationality (in either sense) than did those of von Neumann and Morgenstern.

However, Nash continues to impose stringent limitations on the threats that the agents may choose. Because threats can only influence a disagreement point that is otherwise given, Nash implicitly excludes conditional threats such as "If you do not grant me a final payoff of at least 12, then I will adopt strategy S_2, but otherwise S_1." Such a contingent threat, if credible, would make the disagreement point depend on the players' offers, while for Nash's theory they do not. It could, at the least, introduce discontinuities and non-convexities into the utility possibility frontier. If the contingent threats are consistent, as we might expect for perfectly informed players, then the utility possibility frontier could be reduced to a point, and the final bargain determined strictly by the threats, not by the second-stage maximization. On the other side, it is not at all clear why a rational person would foreswear conditional threats if threats are enforceable.

Nash's variable-threat bargaining does seem to be an advance relative to assumption AV, since it will apply to cases in which the most hostile possible threat does not in fact enhance the threatener's bargaining position, which seems to be a real possibility. (This advance does not seem to have been widely noticed.) However, the limits Nash places on threat behavior remain questionable, if players are indeed rational and ruthlessly self-interested.

2.5. Nash Bargaining and Nash Equilibrium

It seems to be widely believed (1) that cooperative game theory can be applied where agreements are enforceable and (2) that the Nash bargaining solution, as the application of Nash equilibrium to a bargaining game, is implied by individual rationality. Neither of these widespread beliefs is true. We have already seen that enforceability of threats as well as agreements is necessary, if individuals are not cooperatively rational. Moreover, as we have seen, there is at best some tension between individual rationality and the assumptions of both Nash's bargaining theories. Further, a careful reading of Nash (1953) shows that (2) is fatally oversimplified.[6]

It is of course true that Nash offers his well-known "demand game" as a noncooperative model of bargaining that yields the same solution as

---

[6]Rubinstein's (1982) paper is also referenced in support of this belief. However, Rubinstein's model reduces to the Nash solution only in a limiting special case, and corresponds more closely to Cross' (1975) model.

the axiomatically derived theory. In the demand game, (once the disagreement outcome is determined by the threat strategies) each player states
his minimum demand. If the demanded pair $\{u, v\}$ is feasible, then the
two players get u and v, respectively; otherwise the disagreement outcome.
But the Nash bargaining solution is not *generically* a Nash equilibrium
of this game. As Nash notes, "The demand game defined by these payoff
functions will generally have an infinite number of inequivalent equilibrium
points" (Nash, 1953, p. 131). Indeed, given the compactness of the utility
possibility set, any individually rational Pareto-optimal payoff pair will be
a Nash equilibrium of the demand game. To resolve this, Nash presents
what clearly seems to be the first refinement of the Nash equilibrium. He
introduces a continuous *smoothing function* h that "can be thought of as
representing uncertainties in the information structure of the game, the
utility scales, etc" (Nash, 1953, p. 132). This function is suppose to be 1 on
the actual utility possibility frontier and to approach zero for vectors u, v
that are beyond it. As we move outside the utility possibility frontier, the
smoothing function must approach zero in a very regular way that Nash
does not make explicit. Then, if h is sufficiently regular, the Nash equilibrium of the demand game over the smoothed utility possibility frontier will
be unique. We then consider a sequence of Nash equilibria as the game is
played over a sequence of smoothing functions so that h approaches the
discontinuous characteristic function of the utility possibility set. A particular Nash equilibrium of the original demand game will be the limit of this
sequence, and that limiting equilibrium corresponds to the solution from
the axiomatic derivation.

For a specific example, let Game 2.3 be defined by the utility possibility
frontier S together with disagreement point $(0, 0)$, where

$$S = \{(u, v) | f(u, v) \leq 0\} \tag{2.7}$$

where f is a continuous, concave and continuously differentiable function
increasing in u and v and

$$f(0, 0) < 0. \tag{2.8}$$

The characteristic function is

$$g(u, v) = \begin{cases} 1 & \text{if } f(u, v) \leq 0 \\ 0 & \text{if } f(u, v) > 0 \end{cases} \tag{2.9}$$

let

$$\theta = \max(1 - kf(u, v), 0) \tag{2.10}$$

where $k > 0$. The "smoothed" characteristic function for the utility possibility frontier might be

$$h(u, v) = \begin{cases} 1 & \text{if } f(u, v) \leq 0 \\ \theta & \text{if } f(u, v) > 0 \end{cases}. \tag{2.11}$$

Now, if bargainer a demands u, his expected value payoff, taking v as given, will be

$$E(u) = \theta u. \tag{2.12}$$

A necessary condition for a maximum of $E(u)$ is

$$u\frac{\partial f}{\partial v} = \frac{1}{k} - f(u, v). \tag{2.13}$$

Note that if the left-hand side is positive, necessarily $f(u, v) < \frac{1}{k}$. Moreover, symmetrically,

$$v\frac{\partial f}{\partial v} = \frac{1}{k} - f(u, v) \tag{2.14}$$

so that

$$u\frac{\partial f}{\partial u} = v\frac{\partial f}{\partial v}. \tag{2.15}$$

Now let k increase without bound. Clearly, $f(u, v) \to 0$. Since (2.15) applies for every k, we have a sequence of equilibria in the smoothed games that approaches the particular Nash equilibrium at which (2.15) applies, and that is the Nash bargaining solution for this game.

The function θ might, as Nash suggests, be thought of as expressive of agents' common subjective estimates that the vector (u, v) is feasible. For this example, the agents always overestimate, but never underestimate the extent of the feasible region; however, this could probably be remedied. When Nash proposed this approach, it was the first proposal to obtain a unique equilibrium via refinement, but today there are roughly as many proposed refinements of Nash equilibrium as there are cooperative game solution concepts. Following Schelling's (e.g., 1960) line of thought, we might suggest that refinements of Nash equilibrium miss the point. When we have a game that seems to describe some real world interaction, and it has many Nash equilibria, that simply is telling us that there may be many

situations in the real world that can be consistent with individual rationality in the noncooperative sense. And we may see different equilibria in
different observed instances of the same interaction. (McCain, 2010, p. 85)

We see, in any case, that the Nash bargaining solution is not a necessary
condition for a Nash equilibrium even in a simplified bargaining game such
as the Demand Game. Rather, in addition to the Nash equilibrium, we must
make further assumptions that are not derived from individual rationality,
and in the process reconsider the assumption of perfect knowledge about
the structure of the game. Conversely, any Pareto-optimal imputation is a
Nash equilibrium in the Demand Game.

2.6. Unequal Bargaining Power

In Nash (1950, p. 153) equal bargaining power is explicitly assumed, but
in Nash (1953, p. 138), the assumption is retracted and it is argued
instead that the concept of bargaining power has no proper application
to bargaining among rational agents. Nevertheless, pragmatically, it may
be useful to assume differing bargaining power. Consider the business firm,
for example. On one hand, the institutions of a capitalist economy may
result in the employer having disproportionate bargaining power *vis-à-vis*
employees. On the other hand, if the employees join together and have
specialized skills, such as harbor pilots or football players, this may seem
to shift the bargaining power strongly in their direction. Undoubtedly, it
would be good to model the moves in the underlying game that would
produce those results. However, where we are mainly interested in patterns
in the division of the benefits among a cooperating group, i.e., in value solutions, this may be infeasible and we may instead use a solution concept that
allows arbitrary differences in bargaining power. In the 1970s and 1980s,
this was addressed especially by Roth (1979) and Svejnar (1986).

Roth addresses a somewhat more general case, the n-person "bargaining
problem." This is a special class of cooperative games in which no coalition
other than the grand coalition can obtain any benefit from cooperation.
That is, letting N be the set of players and $C \subset N$, then the utility possibility frontier for C is a single point, the disagreement outcome. For a
two-person game, of course, this restriction makes no difference. Roth then
reprises Nash's proof and explores slightly weaker conditions that generate
the same result, and then explores a generalization along the following lines.
Let $\{\alpha_i\}$ be a strictly positive vector of payoffs that is Pareto optimal given

the utility possibility frontier for a game Γ. In place of Nash's symmetry assumption, Roth adopts the more general axiom that if Γ is a symmetrical game, the solution is the vector $\{\alpha_i\}$. The solution of this more general case for any bargaining problem Γ is then the vector that maximizes the weighted geometric mean of the net payoffs $(u_i - d_i)$, that is,

$$\max \prod_{i \in N} (u_i - d_i)^{\alpha_i} \quad \text{subject to } \{u_i\} \in S. \tag{2.16}$$

Roth remarks (1979, p. 17) that if $\alpha \neq \{1, 1, \ldots, 1\}$,[7] "then we have some information that the bargaining abilities of the players... are not equal." Roth (1979, p. 19) goes on to note that complications may arise when $|N| > 2$ that do not arise in two-person bargaining games.

Once again we can illustrate this solution for Game 2.1. Suppose that individual 2 has bargaining power 0.667 and individual 1 has 0.333. Then we have the solution shown in Fig. 2.3. The hyperbola represents a constant value of $(u - d_u)^{0.333}(v - d_v)^{0.667}$ equal to 7.31829, and we see that the maximum corresponds to a cooperative choice of the pure strategy pair $\{$up, left$\}$ and equal payoffs of 10, 10. The increase of the second bargainer's bargaining power from 0.5 to 0.667 results in a shift of the solution from the mixed strategy with probability of 0.9261 for $\{$up, left$\}$ to the pure strategy, i.e., a probability of 1. The shift might have been greater but for the corner solution. In this case, the increase of the second bargainer's bargaining power has just offset the asymmetrical advantage that the underlying game in normal form gives to the first bargainer.

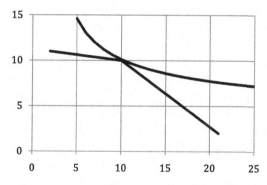

Fig. 2.3. Roth's solution to Game 2.1 if bargainer 2 has twice the bargaining power of bargainer 1.

---

[7]For consistency, $\alpha \neq \{\frac{1}{n}, \frac{1}{n}, \ldots, \frac{1}{n}\}$, but there is no substantive difference.

The asymmetrical model has strong intuitive appeal, given the assumptions of independence of linear transformations of the utility function and of irrelevant alternatives. If we suppose that bargainers may have different bargaining capabilities, the payoffs they would obtain in a symmetrical game provide a natural index of their bargaining power. Roth suggests in a footnote (1979, p. 17) that the asymmetric extension of Nash equilibrium was well established as a conjecture in the oral tradition of game theory.

Svejnar (1986) presents a theoretical model of bargaining along similar lines[8] and uses it to estimate the bargaining power of some American unions. Svejnar's model differs from that of Roth in that he allows the bargaining power to be a function of a vector of institutional parameters. This enables him to test some hypotheses from institutional writing on bargaining. The hypothesis of equal bargaining power, i.e., the original Nash solution, is also testable in this approach. As with Roth, Svejnar's axiomatic derivation addresses an n-person bargaining problem. Svejnar follows Nash and Roth in adopting the two independence axioms, substitutes for Pareto optimality the assumption that the solution is strictly Pareto preferable to the disagreement point, and adopts from Aumann and Kurz (1977) a concept of "fear of disagreement."

Aumann and Kurz had written about the determination of taxes in a simplified majority voting game, which is not a bargaining problem; and they assume a continuum of voters. They adapt Shapley's (1969) procedure for a value solution for an NTU game. The utility possibility frontier is determined by a set of individual endowments e_i that can be redistributed or destroyed, and a set of von Neumann–Mogenstern utility indices u_i defined as functions of the endowment net of the portion redistributed or destroyed. To determine the value of a coalition, a vector $\{\lambda_i\}$ of weights applicable to the utility indices will be required. For the Aumann and Kurz game, this determines only the value of the grand coalition, an aggregate utility $U = \sum_{i \in N} \lambda_i u_i$. For $C \subset N$, the value $v(C)$ is determined by a Nash variable-threat bargaining game. The threats are threats to destroy part of one's endowment, thus reducing the aggregate wealth and utility. The reduced game over the threats is solved by a min-max operation which yields an even division of the aggregate utility between C and $N \backslash C$. However, this even division of *utility* may not be feasible since it may require a redistribution of utility. Following Shapley, Aumann and Kurz (1977) argue that *for their game* there is exactly one vector $\{\lambda_i\}$ that yields an allocation

---

[8]Svejnar's conception was at least partly independent of Roth's. I recall seeing tentative drafts before 1979.

of utility that can be attained without redistribution of utility and so is feasible. The corresponding allocation of utility (and net income) is the value solution they adopt (Aumann and Kurz, 1997; pp. 1143–1145). This corresponds to Shapley's (1969) value solution for NTU games; for these games the uniqueness of the solution cannot be assured in general,[9] and in footnotes, Aumann and Kurz remark that they will "steadfastly ignore" fine points such as the existence of the maximum of aggregate utility, the applicability of the minimax theorem, and the uniqueness of λ for any larger class of games (Aumann and Kurz, 1997; pp. 1143–1145, notes 12–14). At the second stage of the variable-threat bargaining game, the ratio of individual i's utility to his marginal utility of income, the "fear of ruin," $\frac{u_i}{u_i'}$, plays a key part (Aumann and Kurz, 1997; pp. 1147–1149). Thus, they observe that i's "income tax $= \frac{u_i}{u_i'} - c$, where c is a constant." Their discussion on this point recalls that of Zeuthen.

Svejnar uses instead the term "fear of disagreement." He incorporates this ratio in the axiom he posits that replaces Nash's symmetry axiom (Svejnar, 1986; p. 1061). This limits the scope of his solution to a case in which the individuals derive their utility from the division of a given total "money" surplus, a case with many economic applications. Thus, in one sense Svejnar's bargaining theory is not quite as general as the theorems of Nash and Roth. In another sense, though, it is more general in that it allows for risk aversion, which plays a key part.

Svejnar's axiom states that, for a symmetrical game, at the solution, the ratio of fear of disagreement to bargaining power is the same for all bargainers. This leads quite directly to the result that the bargain will correspond to the maximum at Eq. (2.16) (Svejnar, 1986; p. 1063). Svejnar also interprets the axiom in terms of the bargaining process in terms similar to Zeuthen (1930, p. 1062) and it seems fair to say that Svejnar's discussion bridges any remaining gap between the Nash–Roth and Zeuthen approaches.

Svejnar then uses this model to estimate the bargaining power and risk aversion of unions bargaining with a number of major companies in the period approximately from 1950 to 1980. His estimates of union bargaining power range widely, from 0.15 to 0.85, where, for bargaining power defined as he does, equal bargaining power would correspond to a union power index of $\frac{1}{2}$. (Svejnar, 1986; p. 1071; note also pp. 1073, 1074). It seems that

---

[9]This is at page 11 in the preliminary working paper for the RAND corporation, Shapley (1967). Available at: http://www.rand.org/pubs/papers/2008/P3582.pdf. Accessed 24 June 2011.

the symmetry axiom can be empirically rejected, at least for American collective bargaining in the third quarter of the 20th century.

The extensions by Roth and Svejnar are valuable ones. The extension to an n-person bargaining problem is more limited, since few n-person games can be reduced to the bargaining problem, except through some *ad hoc* aggregation. However, for two-person games, these extensions give strong formal and empirical support to an intuitively appealing model for games with unequal bargaining power, and this seems a conceptually and empirically important case.

2.7.   Chapter Summary

Bargaining theory addresses problems of interactive or collective decision where all must agree in order to obtain any mutual benefit, and no coalition of less than the whole can obtain any better payoffs than the individuals can get for themselves if there is no agreement. As such, it is a special case for $n > 2$, but it is applicable to all cooperative games with $n = 2$ and thus has usually been applied to this case. Following the work of Zeuthen (1930), there was a body of bargaining theory outside the tradition of game theory, but Nash's studies (1950, 1953) have influenced further developments both within and outside game theory. Nash's theory provides both an axiomatic basis for the solution and a solution that is computationally tractable (and the same as that of Zeuthen); in addition, he provides a noncooperative model in the form of a refinement of Nash equilibrium for a stylized noncooperative bargaining game. These advantages certainly account for the wide influence of Nash's bargaining theory. However, one of Nash's axiomata, the independence of irrelevant alternatives, is problematic, and his assumption that bargaining power is either equal or irrelevant to bargaining theory seems to be empirically false. Extensions to cases of variable bargaining power have come from Roth and Svejnar. They too lead to a natural and tractable computational solution, a generalization of that of Zeuthen and Nash, one with more degrees of freedom and that fits some empirical data better. Despite the widespread impression that the Nash bargaining solution or its generalization is a necessary consequence of individual rationality, it is not. All of the variants of Nash's theory (like other value solutions for cooperative games) rest on assumptions about the rationality *of the solution*, which are either normative or debatable, and usually both.

Chapter 3

Nontransferable Utility Games and Games in Partition Function Form

The purpose of this chapter is to outline some more general concepts and examples of cooperative games: transferable utility (TU) games in partition function form and nontransferable utility (NTU) games in partition function form. These concepts are not new, taken one by one, and much of the material is expository, so far as TU games are concerned. Some of the details of definitions, examples, and some of the critical discussion will be novel. There seems little or no discussion of NTU games in partition function form in the literature, however, so the discussion of Section 3.3 would mostly be novel. We begin by exploring some ways in which the models in this and some future chapters differ from those especially in Chapter 1.

3.1. Preliminaria

As observed in Chapter 2, the predominant traditions in both cooperative and noncooperative game theory are based on extreme assumptions about rationality or about the enforceability of contracts. For the balance of this book, the objective will be to reconsider some of the value solutions and related concepts on the basis of somewhat less extreme assumptions. In place of Assumption CCR we will adopt.

Assumption CR (Cooperative rationality). If an agreement is to the advantage of all those who are party to it, net of the cost of coordinating and enforcing the agreement, then credible commitments to carry out the agreement can be made and threats to make and carry out such a commitment are credible.

We should stress that, like CCR, AV, and TU, CR is not an axiom but a guiding principle. Further, CR is not necessarily consistent with AV. In this connection, consider again Game 1.1 in Chapter 1. In negotiation within the grand coalition, Player a might threaten to withdraw from the coalition and go alone as a singleton, and this threat would (according to Assumption CR) be credible. However, under Assumption CCR, the compound threat to withdraw from the grand coalition *and adopt strategy "down"* is credible, and Assumption AV follows from it. Under Assumption CR, however, the compound threat may or may not be credible. Assumption CR makes no assertion either way. For this chapter, we will adopt a contrary assumption:

Assumption CA. As between distinct coalitions, threats and promises are credible only if they are self-enforcing.

CA stands for "Coalitional Autonomy." Assumption CA then would imply that the compound threat to act independently and adopt strategy "down" is not credible in the game described. Like Nash, we are assuming that a threat will not be carried out unless it were something the coalition "would want to do, just for itself," or were enforced. Suppose that Player 1 were to "compel" himself to carry out the threat of strategy "down" by posting a bond. The move of posting a bond would be a move in a larger game in which Game 1.1 would be embedded, and moreover, would involve the formation of a coalition between Player 1 and the third party who holds the bond. As such, it would be credible under Assumption CR.

If, following Nash, we think in terms of enforceability rather than rationality, we could say that all contracts and threats made for the formation of coalitions are enforced, but other than that no threats are enforced, and that contracts for the formation of coalitions are always "contracts at will," like typical employment contracts and some other contracts in the actual economy.[1] On the whole, Assumptions CR and CA tell us that interactions within coalitions are cooperative, but that interactions between coalitions are noncooperative.

To illustrate the approach, we may consider a simple public goods provision game, Game 3.1. We consider a three-person game in which the participants are A, B, and C. Each begins the game with wealth of 5 and may choose to produce, or not to produce, one unit of a public good that will increase the wealth of the three agents each by one; but the agent who

---

[1]However, compare McCain (2009, pp. 178, 189, Chapter 10, Section 4), and for contrasting, and more complicated, views.

Table 3.1.   Game 3.1: A public goods game.

Payoffs: a, b, c	C			
	Produce		Don't	
	b		b	
	Produce	Don't	Produce	Don't
a Produce	6.5, 6.5, 6.5	5.5, 7, 5.5	5.5, 5.5, 7	5.5, 6,6
Don't	7, 5.5, 5.5	6, 6, 4.5	6, 4.5, 6	5, 5, 5

produces a public good undergoes a private cost of 1.5 units of wealth. For an agent who does not produce the public good, payoffs if zero, one, or two units are produced are thus 5, 6, or 7; for an agent who does produce, if the quantity of public good produced is 1, 2, or 3 units the payoffs are 4.5, 5.5, and 6.5. Thus the game is as shown in Table 3.1.

Clearly, the value of the grand coalition is 19.5 and the value of a two-person coalition is 11. But what about the value of a singleton? This may depend on whether the other two form a coalition or not. According to Assumption AV, the two-person coalition would threaten to refuse to produce any of the public good, reducing the value of the holdout to 5 at the cost of reducing the value of their own coalition to 10. However, Assumption CA excludes any threats, promises, or side payments *among coalitions*, so that the two-person coalition will choose their dominant strategy: produce two units of the public good. Thus the value of the singleton will be 7. On the other hand, if the two-person coalition does not form, then the value of the singleton is 5.

In a three-person game, as we see, the value of a singleton may depend on whether the group as a whole is partitioned as 2, 1 or 1, 1, 1, supposing that the play among the coalitions is noncooperative. More generally, in an n-person game, the value of any coalition with a membership of $n - 2$ or less will depend on the partition. That is, it will be a game in *partition function form*:

Definition 3.1. A game in partition function form is a pair $\{N, v(\mathcal{P}, C)\}$, where N is a set of agents, $\mathcal{P}$ a partition of N, $C \in \mathcal{P}$, and v a function that determines the value of the coalition C as embedded in the partition $\mathcal{P}$.

Games in partition function form have been used to represent externalities (e.g., Ray and Vohra, 1999; Carraro, 2003; Pintassilgo and Lindroos, 2008).

Telser argues (1978, esp. p. 13, note also McCain, 2009, pp. 153–154) that, for a game in coalition function form, Assumption AV is the more appropriate one *because* the Nash equilibrium will in general depend on the partition, while the assurance value does not. Conversely, if we choose the partition function form, Assumption CA seems the more appropriate one. The appropriate basis for the choice is pragmatic, and the games in partition function form allow us to model externalities and holdout behavior explicitly, as Assumption AV does not.

Assumptions CR and CA are consistent, in principle, with superadditivity. However, superadditivity will not be assumed in general. The next section expands on the representation of TU games in partition function form.

3.2. TU Games in Partition Function Form

As we have seen in Game 3.1, when there are public goods or externalities, if we allow for free rider or holdout behavior, the value of a coalition may depend on the partition in which it is embedded. The TU game in partition function form dates from a paper of Thrall and Lucas (1963), but has gained in interest since the 1990s, largely through interest in problems of externalities.

Let N be an index set of agents in a game, $i\varepsilon N, i = 1, \ldots, n$, and let Π_N be the set of all partitions of N and $\mathcal{P} \in \Pi_N$. $\mathcal{P} = \{C_1, C_2, \ldots, C_r, \varnothing\}$. $|C_i|$ will denote the number of members in C_i and $|\mathcal{P}|$ will denote the number of nonempty coalitions in $\mathcal{P}$. A pair $\{\mathcal{P}, C_i\}$ with $C_i \in \mathcal{P}$ is called an embedded coalition. A partition function $v(\mathcal{P}, C_i)$ assigns a real number value to coalition C_i in the context of the partition $\mathcal{P}$. $\Gamma = \{N, v(\mathcal{P}, C_i)\}$ comprises a TU game in partition function form. In the balance of this section, $\Gamma(N, v)$ is a TU game in partition function form, $\mathcal{P} \in \Pi_N$, and $|\mathcal{P}| = r$, i.e., $\mathcal{P}$ has r nonempty coalitions. In this chapter, from this point on, TU games will be represented in partition function form.

If $\mathcal{Q} = \{B_1, \ldots, B_s, \varnothing\} \in \Pi_N$, and $\forall i = 1, \ldots, s, B_i \neq \varnothing \Rightarrow \exists k \in \{1, \ldots, r\} \ni \forall B_i \subseteq C_k$, then $\mathcal{Q}$ is said to be a *refinement* of $\mathcal{P}$. The *fine* partition is $\mathcal{F} = \{\{a_1\}, \{a_2\}, \ldots, \{a_n\}\}$.

Lemma 3.1. $\forall \mathcal{P} \in \Pi_N, \mathcal{P} \neq \mathcal{F}, \mathcal{F}$ *is a refinement of* $\mathcal{P}$.

Let $\mathcal{G} = \{N, \varnothing\}$, the partition comprising only the grand coalition.

Lemma 3.2. $\forall \mathcal{P} \in \Pi_N, \mathcal{P} \neq \mathcal{G}, \mathcal{P}$ *is a refinement of* $\mathcal{G}$.

Remark. These lemmas follow trivially from the definitions.

Definition 3.2. Let $\mathbf{x}$ be a point in $\Re^N$, and $x_C = \sum_{i \in C} x_i$. Then $\mathbf{x}$ is an imputation admissible for partition $\mathcal{P}$ iff $\forall C \in \mathcal{P}, x_C = v(\mathcal{P}, C)$.

Remarks: It is conventional in the theory of superadditive TU games in coalition function form to distinguish between a pre-imputation and an imputation, where a pre-imputation is, in the terms of this definition, admissible for the grand coalition, and an imputation satisfies the further condition of individual rationality, i.e., for all $i, x_i \geq v(\mathcal{F}, \{i\})$. (See, e.g., Forgo *et al.*, 1999, p. 222; Peleg and Sudhölter, 2003, p. 26.) This presupposes that the grand coalition is efficient and will ultimately be formed, but because this chapter does not adopt that presupposition, this terminology is not helpful and the above definition is substituted. In assuming that the payoffs for a coalition can be no more than its value, this definition follows Aumann and Dreze (1974), although much subsequent literature has not followed their approach. It is also the terminology of McCain (2009, Chap. 13).

The principal concern of this book, as the title indicates, is value solutions, i.e., with imputations that can be distinguished by some conditions that they uniquely satisfy. Since admissible imputations depend on the partitions, ideally the solution will specify the partition as well as the imputation. Thus a candidate solution will be a pair, $\mathcal{P}, \mathbf{x}$, with $\mathbf{x}$ admissible for $\mathcal{P}$. If this is not possible, then we may consider as a solution a list of imputations conditional on partitions, such that in each case the imputation uniquely satisfies the conditions for that partition.

3.2.1. *Symmetry*

Let $\Gamma = \{N, v(\mathcal{P}, C)\}$ and let $\pi(N)$ be a permutation of N. For each $\mathcal{P} \in \Pi$, construct $\mathcal{Q}$ so that, for $C \in \mathcal{P}, C = \{i, j, \ldots, k\}, B = \{\pi(i), \pi(j), \ldots, \pi(k)\} \Rightarrow B \in \mathcal{Q}$.

Lemma 3.3. $\mathcal{Q} \in \Pi_N$.

Remark. Again, Lemma 3.3 follows trivially from the definition.

For $C \in \mathcal{P}, C = \{i, j, \ldots, k\}, B = \{\pi(i), \pi(j), \ldots, \pi(k)\}$, denote B as $\pi(C)$.

Definition 3.3. Iff $v(\mathcal{P}, C) = v(\mathcal{Q}, \pi(C))$ for all $C \in \mathcal{P}, \Gamma$ is symmetrical.

Definition 3.4. Let Γ be a symmetrical game in partition function form, and φ be a value solution for Γ. Suppose that, $\forall S \in \mathcal{P} \in \Pi_N, s = |S|, i \in S, \varphi_i(G, \mathcal{P}) = \frac{v(\mathcal{P},S)}{s}$. Then φ is a *symmetrical* value solution, or put otherwise, φ has the *symmetry property*.

Remark. The symmetry property expresses a judgment or condition that there are no exogenous differences in bargaining power. Endogenous differences in bargaining power might arise from unsymmetrical competitive alternatives, but in a game that is symmetrical in this sense, all agents have similar competitive alternatives and so, in the absence of exogenous differences in bargaining power, equal division of the value of the coalition is to be expected.

3.2.2. *Superadditivity*

To define superadditivity for a partition function game calls for a little care.

Definition 3.5. Suppose that

(a) $\mathcal{P}$ is a refinement of $\mathcal{Q}$.
(b) $C_1 \in \mathcal{Q}, C_2 \in \mathcal{P}, C_3 \in \mathcal{P}$, and $C_1 = C_2 \cup C_3$.
(c) $C_4 \in \mathcal{Q}, C_4 \neq C_1 \Rightarrow \exists C_5 \in \mathcal{P} \ni C_4 = C_5$.
(d) $C_6 \in \mathcal{P}, C_6 \neq C_2, C_6 \neq C_3 \Rightarrow \exists C_7 \in \mathcal{Q} \ni C_6 = C_7$.

Iff v is such that: (a)–(d) $\Rightarrow v(\mathcal{Q}, C_1) \geq v(\mathcal{P}, C_2) + v(\mathcal{P}, C_3)$, then Γ is superadditive.

Remark: Item (b) asserts that C_1 is formed by the merger of C_2 and C_3, and (a), (c), and (d) assert that apart from those coalitions, the two partitions are identical. Without (a), (c), and (d), it might be that negative externalities from the reorganization of agents not in the merger would reduce the value of the merged coalition so that the inequality would not be observed.[2]

Definition 3.6. Let $\mathbf{x}$ be an imputation admissible for $\mathcal{P}, C \subseteq N, C \notin \mathcal{P}$. Suppose $\exists \mathcal{Q} \in \Pi_N \ni C \in \mathcal{Q}$ and $v(\mathcal{Q}, C) \geq \sum_{i \in C} x_i$, then $\mathcal{Q}$ dominates $\{\mathbf{x}, \mathcal{P}\}$ via C.

Remarks: This is the conventional definition of dominance adapted to partition function games. If we have $v(\mathcal{Q}, C) = \sum_{i \in C} x_i$, then the dominance is weak; otherwise it is strict.

---

[2]For a more detailed discussion see McCain (2009, pp. 195–197).

Table 3.2. Game 3.2.

Partition	Values
$\{a, b, c\}$	35
$\{a, b\}, \{c\}$	30, 1
$\{a, c\}, \{b\}$	30, 1
$\{b, c\}, \{a\}$	30, 1
$\{a\}, \{b\}, \{c\}$	5, 5, 5

Definition 3.7. Suppose $\exists C \subseteq N$ and $\exists \mathcal{Q} \in \Pi_N \ni C \in \mathcal{Q}$ and for any $\mathbf{x}$ admissible for $\mathcal{P}$, $\mathcal{Q}$ dominates $\{\mathbf{x}, \mathcal{P}\}$ via C. Then $\mathcal{Q}$ unconditionally dominates (u. d.) $\mathcal{P}$ via C.

Remarks. In one sense, unconditional dominance is a stronger condition, since unconditional dominance will imply dominance for any admissible imputation. However, in another sense, it is a weaker condition, in that the set of partitions that are not unconditionally dominated will be a larger set than the set of all partitions for which there are no undominated imputations. Consider Game 3.2, as shown in Table 3.2. For this game in partition function form, the grand coalition is dominated by one or another of the 2×1 games, for any admissible imputation; but the grand coalition is not unconditionally dominated by any other coalition.

Lemma 3.4. *If Γ is superadditive, and $\mathcal{P}, \mathcal{Q}$ are as (a)–(d) above, then $\mathcal{Q}$ u. d. $\mathcal{P}$ via C_1.*

Lemma 3.5. *If for any $\mathcal{P}, \mathcal{Q}$ characterized by (a)–(d) above, $\mathcal{Q}$ u. d. $\mathcal{P}$ via C_1, then Γ is superadditive.*

Remarks. The first follows because, for imputations admissible for $\mathcal{P}$, x_{C1} can be no more than $v(\mathcal{P}, C_2) + v(\mathcal{P}, C_3)$. The second essentially restates the definition of superadditivity. We see that unconditional dominance is closely related to superadditivity.

3.2.3. *Efficiency*

The usual definition of efficiency in economics is Paretian, and this can readily be adapted to games in partition function form.

Definition 3.8. Let $\mathcal{P} \in \Pi_N, \mathcal{Q} \in \Pi_N, \mathbf{x}$ admissible for $\mathcal{P}$, $\mathbf{y}$ admissible for $\mathcal{Q}$, and $\forall i \in N, y_i \geq x_i$; moreover, $\exists i \in N \ni y_i > x_i$. Then $\mathcal{Q}, \mathbf{y}$ is Pareto-superior (P-s) to $\mathcal{P}, \mathbf{x}$.

Definition 3.9. Suppose that $\mathcal{P}, \mathbf{x}$ is such that for $\mathbf{y}$ admissible for $\mathcal{Q} \in \Pi_N, \exists i \in N \ni y_i > x_i \Rightarrow \exists j \in N \ni y_j < x_j$. That is, there is no candidate solution that is Pareto-superior to $\mathcal{P}, \mathbf{x}$. Then $\mathcal{P}, \mathbf{x}$ is efficient in the Paretian sense, that is, Pareto-optimal.

Remark. Note that we have *not* assumed that $\mathcal{P} \neq \mathcal{Q}$, so $\mathbf{x}$ and $\mathbf{y}$ could be alternative imputations for the same partition.

Definition 3.10. Suppose that, for any $\mathbf{x}$ admissible for $\mathcal{P}, \exists \mathbf{y}$ admissible for $\mathcal{Q} \ni \mathcal{Q}, \mathbf{y}$ P-s $\mathcal{P}, \mathbf{x}$. Then we may say that $\mathcal{Q}$ is Pareto-superior to $\mathcal{P}$.

These concepts (and superadditivity) will be illustrated by Game 3.3, a game with some surprising properties that will accordingly be called the Enigma Game. The Enigma Game is a four-person game shown in partition function form in Table 3.3.

Although this game is superadditive, and indeed strictly superadditive, the fine partition is efficient in the Paretian sense for the only imputation that is admissible for the fine partition, $2, 2, 2, 2$. None of partitions 2-14 is Pareto-efficient with any imputation, since the grand coalition, partition 15, is Pareto-superior to all of them with any imputation admissible for partitions 2-14. There are some imputations for which partition 15 is Pareto-efficient, namely imputations that give at least one agent a payoff

Table 3.3.   Game 3.3: The enigma game: A superadditive game with strong negative externalities.

1	$\{a\}, \{b\}, \{c\}, \{d\}$	2, 2, 2, 2
2	$\{a, b\}, \{c\}, \{d\}$	5, 0, 0
3	$\{a, c\}, \{b\}, \{d\}$	5, 0, 0
4	$\{a\}, \{b, c\}, \{d\}$	0, 5, 0
5	$\{a, b, c\}, \{d\}$	6, 0
6	$\{a, d\}, \{b\}, \{c\}$	5, 0, 0
7	$\{a\}, \{b, d\}, \{c\}$	0, 5, 0
8	$\{a\}, \{b\}, \{c, d\}$	0, 0, 5
9	$\{a, b, d\}, \{c\}$	6, 0
10	$\{a, b\}, \{c, d\}$	5, 1
11	$\{a, c, d\}, \{b\}$	6, 0
12	$\{a, c\} \{b, d\}$	5, 1
13	$\{a, d\} \{b, c\}$	5, 1
14	$\{a\} \{b, c, d\}$	0, 6
15	$\{a, b, c, d\}$	7

greater than 2. Moreover, partition 1 is strictly unconditionally dominated by partitions 2–4, 6–8, 10, 12–14 and non-strictly unconditionally dominated by 5, 9, and 11, although (1) none of these is efficient for any admissible imputation, while partition 1 is Pareto-efficient, and (2) all of these are unconditionally dominated by the grand coalition, partition 15, though it does not dominate partition 1. These odd characteristics of the game are consequences of the strong negative externalities in the game.

It appears that (in the presence of negative externalities, at least) there is little relation between dominance and efficiency for games in partition function form. In fact a very weak theorem can be proved relating them.

Theorem 3.1. *If $\mathcal{P} \in \Pi_N$, $\mathcal{P} \neq \mathcal{G}$ the partition comprising the grand coalition, and $\mathcal{G}$ strictly unconditionally dominates $\mathcal{P}$, then $\mathcal{G}$ is Pareto-superior to $\mathcal{P}$.*

Proof. Since $\mathcal{G} = \{N, \varnothing\}$, it must be that $\mathcal{G}$ dominates $\mathcal{P}$ via N. Thus, for any $\mathbf{x}$ admissible for $\mathcal{P}, v(\mathcal{G}, N) > \sum_{i=1}^{n} x_i$. For $i \neq 1$, let $y_i = x_i$ and let $y_1 = v(\mathcal{G}, N) - \sum_{i=2}^{n} x_i$. Then $\mathcal{G}, y$ is Pareto superior to $\mathcal{P}, \mathbf{x}$. $\square$

3.2.4. *Successor Function*

Let $\Gamma = \{N, v(\mathcal{P}, c)\}, \mathcal{P} \in \Pi_N$, and $C \notin \mathcal{P}$, that is, C is a deviating potential coalition. If C should form, what value could its members expect to realize? For that question, we need to know what partition $\mathcal{Q}$ would follow from the defection of C; and since there may be several $\mathcal{Q}$ such that $C \in \mathcal{Q}$, the value that C can expect to realize may be ambiguous (or worse!) McCain, 2009, assumes an arbitrary successor function.

$$\sigma : \{\mathcal{P}, C\} \to \Pi_N \ni \mathcal{Q} = \sigma(\mathcal{P}, C) \Rightarrow C \in \mathcal{Q}. \tag{3.1}$$

McCain makes some remarks on what might characterize a rational successor function. One advantage of an arbitrary, given successor function is that the results obtained are independent of the particular successor function assumed, whether rational, optimistic, pessimistic, or as Thrall and Lucas assume,

$$\sigma(\mathcal{P}, C) = \mathcal{R} \ni v(\mathcal{R}, C) = \min_{\mathcal{Q} \in \Pi_N \ni C \in \mathcal{Q}} v(\mathcal{Q}, C). \tag{3.2}$$

Expression (3.2) is consistent with Assumption AV.

Definition 3.11. If $\mathcal{P} \in \Pi_N, \mathcal{P} = \{C_1, C_2, \ldots, C_r\}, S \notin \mathcal{P}$, then $\mathcal{P}^S = \{S, C_1 \backslash S, C_2 \backslash S, \ldots, C_r \backslash S\}$.

Definition 3.12. If $\forall \mathcal{P} \in \Pi_N, \forall S \notin \mathcal{P}, \sigma(\mathcal{P}, S) = \mathcal{P}^S$, then σ is the naïve successor function.

For this study, however, we will take a slightly different direction. First, we will limit our attention to successor functions that are invariant for strategically equivalent games. That is, given $\Gamma = (N, v(\mathcal{P}, c))$, consider $\Gamma^* = (N, w(\mathcal{P}, c))$ where $w(\mathcal{P}, c) = a + b\, v(\mathcal{P}, c)$, where b is a positive constant and a is any constant. Following convention, Γ and Γ^* are strategically equivalent games. Then σ is invariant over strategically equivalent games if $[\sigma(\mathcal{P}, c) = \mathcal{R}$ for game $\Gamma] \Leftrightarrow [\sigma(\mathcal{P}, c) = \mathcal{R}$ for game $\Gamma^*]$. To impose invariance over strategically equivalent games is essentially definitional, since it is required by the identification of the payoffs as von Neumann–Morgenstern utilities.

To motivate the comments that follow, consider the four-person game, Game 3.4, shown in partition function form in Table 3.4. Suppose, then, that $\mathcal{P}$ is partition 11, and $\{d\}$ deviates. The immediate result is $\mathcal{P}^S$, partition, 15. But $\{a\}, \{b\}, \{c\}$ might reorganize themselves. In effect $\{a\}, \{b\}, \{c\}$ have to play a new game among themselves, and that new game is Game 3.2, shown in Table 3.2. In that smaller game, any imputation admissible

Table 3.4.   Game 3.4: A four-person symmetrical game.

	Partition	Values
1	$\{a, b, c, d\}$	50
2	$\{a, b\}, \{c, d\}$	20, 20
3	$\{a, c\}, \{b, d\}$	20, 20
4	$\{a, d\}, \{b, c\}$	20, 20
5	$\{a\}, \{b, c, d\}$	15, 35
6	$\{b\}, \{a, c, d\}$	15, 35
7	$\{a, b, d\}, \{c\}$	35, 15
8	$\{a, b, c\}, \{d\}$	35, 15
9	$\{a, b\}, \{c\}, \{d\}$	30, 1, 1
10	$\{a, c\}, \{b\}, \{d\}$	30, 1, 1
11	$\{a, d\}, \{b\}, \{c\}$	30, 1, 1
12	$\{c, d\}, \{a\}, \{b\}$	30, 1, 1
13	$\{b, d\}, \{a\}, \{c\}$	30, 1, 1
14	$\{b, c\}, \{a\}, \{d\}$	30, 1, 1
15	$\{a\}, \{b\}, \{c\}, \{d\}$	5, 5, 5, 5

for partition 15 will be dominated by some imputation for one or another of partitions 8, 9, 10, or 14. However, agent d cannot anticipate which of these partitions is likely to form. Moreover, each of partitions 9, 10, and 14 are dominated by 8 via $\{a, b, c\}$ and each of 9, 10, and 14 is dominated by the other two via the two-person coalition. That is, Game 3.2 displays a dominance cycle among these four partitions. And this will affect d's decision whether or not to defect, if his payout from the partition 11 is between 1 and 15. Clearly the successor function can be problematic in some applications.

Our concern here is with value solutions, so a natural expectation would be that the successor function be consistent with the value function, in the light of individual rationality. Consistency of the successor function is a plausible condition, but it requires essentially that the value solution and the successor function be derived simultaneously. It might be argued instead that the successor function is logically prior to the solution, and reflects different considerations, in that the solution reflects threats that are not in fact carried out, while the reorganization of the partition by agents not involved in an original deviation requires them actually to take action. After a deviation, some cooperative relations remain in place among the remnant, as expressed in the partition $\mathcal{P}^S$. Again referring to Game 3.3, suppose that $\mathcal{P}$ is the grand coalition, partition 1, and consider the deviation of $\{d\}$ from the grand coalition. At $\mathcal{P}^S = \{a, b, c\}$, $\{d\}$, a, b, and c remain in a cooperative group. Is it indeed reasonable to suppose that $\{a, b, c\}$ would then reorganize in a way that makes one of them worse off? McCain (2009, pp. 178, 189) addresses this in terms of an *assumed* contract dynamics, but there might be other, quite different contract dynamics that would prohibit a reorganization that would not be a Pareto improvement for $\{a, b, c\}$. Referring instead to the deviation of $\{d\}$ from partition 11, $\mathcal{P}^S$ is the fine partition, and the noncooperative reorganization among $\{a\}, \{b\}, \{c\}$ seems more plausible, but, as we have seen, it has many noncooperative solutions among which partition 8 is a plausible Schelling focal point. In any case, as we have seen, the consistency condition does not get us very far.

We note that partition 8 unconditionally dominates $\{a\}, \{b\}, \{c\}$ via $\{a, b, c\}$. Now, partition 9 also u. d. $\{a\}, \{b\}, \{c\}$ via $\{a, b\}$. However, partition 8 also u. d. partition 9, so it does not seem likely that partition 9 would be the last step. Further, if (for example) a proposes partition 9 to b, c could counteroffer to one of them based on partition 10 or 14, setting off a recontracting process that has no evident end point. This consideration in itself might deter reasonable agents from making offers of this kind. This suggests that we might define a rational successor function in the following way:

Definition 3.13. Given that $\mathcal{Q} = \sigma(\mathcal{P}, S)$, σ is rational iff

(a) If $\mathcal{Q} \neq \mathcal{P}^S$, then $\exists C \subseteq N \backslash S \ni C \in \mathcal{Q}$, $\mathcal{Q}$ u.d. $\mathcal{P}^S$ *via* C.
(b) If $S \in \mathcal{R} \in \Pi_N$ and u.d. $\exists D \subseteq N \backslash S \ni \mathcal{R}$ u.d. $\mathcal{Q}$ via D, then $\mathcal{R} = \mathcal{Q}$.

Expression (a) says that a deviation will be followed by a reorganization of the residual only if some group within the residual group can benefit from it *regardless* of the imputation in $\mathcal{P}^S$; (b) says that the group will not deviate to a partition that leaves some group necessarily worse than they would be in another partition. For Game 3.3, then, we can identify partition 8 as the rational successor of partition 11 with a deviation by $\{d\}$.

Theorem 3.2. *If* Γ *is superadditive,* σ *rational, then* $\sigma(\mathcal{P}, C) = \{C, N \backslash C\}$.

Proof. By superadditivity, $\{C, N \backslash C\}$ u. d. any $\mathcal{Q} \ni C \in \mathcal{Q}$, so part (a) is fulfilled. Now, suppose $\mathcal{R}$ u.d. $\mathcal{Q}$ via $S \subset N \backslash C$. It must follow that $v(\mathcal{R}, S) \geq v(\mathcal{Q}, N \backslash C)$. If, then, $S \neq N \backslash C$, by superadditivity $v(\mathcal{R}, N \backslash (C \cup S)) \mid v(\mathcal{R}, S) \leq v(\mathcal{Q}, N \backslash C)$. Without loss of generality, however, $v(\mathcal{R}, N \backslash (C \cup S)) > 0$. There is no loss of generality because invariance over strategically equivalent games is assumed, so that we may add enough to each embedded coalition value so that the inequality is correct, and characterize σ for the entire class of strategically equivalent games. However, we now have $v(\mathcal{R}, N \backslash (C \cup S)) < v(\mathcal{Q}, N \backslash C)$, a contradiction. □

If the game is not superadditive, the problem can be more complex. Suppose that, in partition 8 in Game 3.3, the value of 35 were replaced by 25. This would violate superadditivity. Letting $\mathcal{P}$ be the grand coalition, partition 1, and again suppose that $\{d\}$ deviates. Then $\mathcal{P}^{\{d\}}$ is partition 8, and it is u.d. by partitions 9, 10, and 14 via the two-person coalitions. Again, however, we might suppose that the proposal of any one of those partitions would begin a process of recontracting that has no evident endpoint, and so we might again exclude such offers. In this spirit we define a surrational successor function as follows:

Definition 3.14. Given that $\mathcal{Q} = \sigma(\mathcal{P}, S)$, σ is surrational iff

(a) σ is rational.
(h) If $\mathcal{Q} \neq \mathcal{P}^S$, $S \in \mathcal{R} \in \Pi_N$ and $\exists D \subseteq N \backslash S \ni D \in \mathcal{R}$, $\mathcal{R}$ u.d. $\mathcal{P}^S$ via D, then $\mathcal{R} = \mathcal{Q}$.

Lemma 3.6. *A surrational successor function exists for any* $\mathcal{P} \in \Pi$ *and is unique.*

Proof. First, clearly $\mathcal{P}^S$ exists for any $\mathcal{P}, S$, and conditions and is the surrational successor unless another partition exists that fulfills conditions (a) and (b) under Definition 3.13 and condition (b) under Definition 3.14. Uniqueness is also trivial because condition (b) in Definition 3.14 above essentially imposes uniqueness of the successor as a condition of surrationality. In effect, these conceptions of rationality and surrationality privilege the status quo following a deviation, since further reorganizations will take place only if some group clearly benefits from them, and no possibility of recontracting cycles exists. In this book, a surrational successor function will be assumed except where it is stated otherwise.

In the following section we consider the extension of the partition function form to NTU games. $\quad\square$

3.3. NTU Games in Partition Function Form

For the purposes of this study, an NTU game in partition function form will be characterized as follows. As before, the "players in the game" are an index set $N = \{1, 2, \ldots, n\}$ and $\mathcal{P} = \{C_1, C_2, \ldots, C_r, \varnothing\}$ is a partition of N. The set of all partitions of N is denoted by Π_N. Let $C, \mathcal{P}$ be an embedded coalition; thus $C \in \mathcal{P}$. Let $C = \{i, j, \ldots, k\} \subseteq N, |C| = m$, with U_i the utility index for agent i. Then the set of feasible utility vectors corresponding to $C, \mathcal{P}$ will be a compact, convex subset of R^m characterized by a function $f_{C,\mathcal{P}}$ such that, for every feasible vector $\{U_i, U_j, \ldots, U_k\}$, $f_{C,\mathcal{P}}(U_i, U_j, \ldots, U_k) \leq 0$. That is, the set of feasible utility vectors is a utility possibility set, in the terms used in welfare economics (Ng, 1980, p. 37). The function $f_{C,\mathcal{P}}$ will be called the *constraint function*. The set characterized by $f_{C,\mathcal{P}}(U_i, U_j, \ldots, U_k) \leq 0$, is called the *utility possibility region*, and its boundary, characterized by $f_{C,\mathcal{P}}(U_i, U_j, \ldots, U_k) = 0$, is called the *utility possibility frontier* for coalition C embedded in partition $\mathcal{P}$. The constraint function can be thought of as an overall index of excess demand for resources. Where it is positive, the utility assignments would require more resources than are available, so that they would not be feasible, but where the constraint function is nonpositive there are sufficient resources available so that the utility assignments are feasible. Denote $U_C = \{U_i\}_{i \in C}$.

Among the properties of a constraint function and a utility possibility set are the following:

(1) $f_{C,\mathcal{P}}(0,0,\ldots,0) < 0$
(2) if $U_i^1 \geq U_i^2, \forall i,$ and $\exists j \ni U_j^1 > U_j^2$ then $f_{C,\mathcal{P}}(U_C^2) < f_{C,\mathcal{P}}(U_C^1);$ and conversely,
(3) if $f_{C,\mathcal{P}}(U_C^2) < f_{C,\mathcal{P}}(U_C^1),$ then $\exists i \ni U_j^1 > U_j^2;$
(4) $\forall i, U_i \geq 0;$
(5) for any two vectors $U_i, U_j, \ldots, U_k$ and $V_i, V_j, \ldots, V_k$ with $f_{C,\mathcal{P}}(U_i, U_j, \ldots, U_k) = 0,$ $f_{C,\mathcal{P}}(V_i, V_j, \ldots, V_k) = 0$ if $U_i < V_i$ then $\exists j \ni U_j > V_j.$
(6) $\exists U_C \ni \forall i, U_i > 0, f_{C,\mathcal{P}}(U_C) < 0.$

Note that, if the utilities are von Neumann–Morgenstern utilities, $U_i > 0$ involves no loss of generality.

Definition 3.15. An NTU game in partition function form will thus be defined as a pair $\{N, \{f_{C,\mathcal{P}}\}\}$, where $f_{C,\mathcal{P}}$ is a constraint function corresponding to each coalition $C \in \mathcal{P} \in \Pi_N$.

We assume that arrangements within the coalition will be Pareto-optimal among its members, so that, for some distributional weights $\{\lambda_i\}$, the U_i's will be such as to (Lange, 1942)

$$\max_{U_i} \sum_{i \in C} \lambda_i U_i \quad \text{subject to } f_{C,\mathcal{P}}(U_C) \leq 0. \tag{3.3}$$

Note that the weights λ_i are defined up to a multiplicative positive transformation, so that without loss of generality $\sum_{i \in C} \lambda_i = k,$ for $k > 0$. The value function for this maximum is

$$V = g(\{\lambda_i\}_{i \in C}), \tag{3.4}$$

a continuous function with a continuous first derivative and

$$U_i = \frac{\partial V}{\partial \lambda_i}. \tag{3.5}$$

Definition 3.16. An imputation U will be said to be admissible for $\mathcal{P}$ (for an NTU game in partition function form) iff $\forall C \in \mathcal{P}, f_{C,\mathcal{P}}(U_C) \leq 0$ and U_C is Pareto-optimal for C.

Remark. It will follow that $f_{C,\mathcal{P}}(U_C) = 0.$

The case of a singleton coalition $C = \{i\}$ deserves some mention. For such a coalition, a compact utility-possibility set is a segment of the real line

$[0, U^{\text{max}}]$. The convexity of this set is trivial. The only Pareto-efficient payment is $U_i = U_i^{\text{max}}$. The utility possibility frontier can be characterized as $f_{c,P}(U_i) = U_i - U_i^{\text{max}} \leq 0$. In particular, for the fine partition $\mathcal{F} = \{\{1\}, \{2\}, \ldots, \{n\}\}$, each singleton coalition has just one efficient imputation, $U_i^{\mathcal{F}}$.

For a TU game in partition function form, the value of a coalition depends on the partition in which the coalition is embedded, so, for example, the merger of two coalitions may modify the value of a third. This is an "externality" in the literature on games in partition function form; and if the value of the third coalition is reduced, it is a negative, and otherwise a positive externality. The discussion of externalities will be somewhat complicated and positive and negative externalities may not always be distinguished for NTU games.

Let $\mathcal{P}$ and $\mathcal{R}$ be two partitions of N, and $A, B, C,$ and D are coalitions with $A \in \mathcal{P}, B \in \mathcal{P}, C = A \cup B$ and $C \in \mathcal{R}$. Thus, one difference between $\mathcal{P}$ and $\mathcal{R}$ is the merger of A and B to form C. Furthermore, $D \in \mathcal{P}$ and $D \neq A, D \neq B \Rightarrow D \in \mathcal{R}$; and $D \in \mathcal{R}$ and $D \neq A \Rightarrow D \in \mathcal{P}$. Suppose then that, for all utility assignments to agents $i \in D, U_C, f_{D,\mathcal{P}}(\{U_i\}) < f_{D,\mathcal{R}}(\{U_i\})$. We may then say that the formation of C has created a negative externality to D. If instead we have $f_{D,\mathcal{P}}(\{U_i\}) > f_{D,\mathcal{R}}(\{U_i\})$, then the externality is positive. But there will be cases in which for some $\{U_i\}, f_{D,\mathcal{P}}(\{U_i\}) < f_{D,\mathcal{R}}(\{U_i\})$, while for other $\{U_j\}, f_{D,\mathcal{P}}(\{U_j\}) > f_{D,\mathcal{R}}(\{U_j\})$. In such a case there is an externality, but its direction is ambiguous.

That is, A and B are merged to form $\mathcal{R}$ from $\mathcal{P}$ and there are no other differences, and suppose that the utility possibility frontier associated with D in $\mathcal{R}$ is shifted inward relative to the utility possibility frontier associated with D in $\mathcal{P}$. Then we may say that there is a negative externality to D from the formation of C. If instead the utility possibility frontier associated with D in $\mathcal{R}$ is shifted outward, then we may say that there is a positive externality to D from the formation of C. It may be, however, that the utility possibility frontier associated with D in $\mathcal{R}$ is rotated, or changed in some more complicated way. In that case we may say that there is an externality from the formation of C to D, but the externality cannot be unambiguously described either as positive or negative. In what follows, we will say that "externalities are absent" in the case for any $\mathcal{P}, \mathcal{R} \in \Pi$э$C \in \mathcal{P}, C \in \mathcal{R}, f_{C,\mathcal{P}} \equiv f_{C,\mathcal{R}}$.

Notice that, thus far, we are imposing the assumption that whenever $i \in C, j \in N \backslash C, \frac{\partial U_i}{\partial U_j} = 0$. This corresponds to an intuition that the productive activity of the coalition (which might create the externality to other coalitions) is determined independently of the distribution of benefits among the members of the coalition, so that the distribution of utility among the members of the coalition reflect only some redistribution of the benefits of the given productive activity, and any nonlinearity in $f_{\mathcal{P},\mathcal{C}}$ reflects the costs of making side payments or the nonlinearity of the utility functions. A more general formulation would associate each partition $\mathcal{P}$ with a utility possibility set $f_{\mathcal{P}}(\{U_i\}_{i \in N})$. This is, however, a trivial extension that will be left to a footnote.[3]

NTU is subject to two interpretations. On one hand, we may have an interpersonally comparable measure of utility; however, practical considerations prevent side payments proportionate to utility. Thus, for a coalition, the total utility value $\sum_{i \in C} U_i$ is meaningful, although it may vary with the distribution of payoffs among the players. We will call this the intermediate case.

Suppose in particular that utility is measureable and

$$f_{C,\mathcal{P}}(U_i, U_j, \dots, U_k) = \sum_{i \in C} U_i - V_{C,\mathcal{P}} \le 0. \qquad (3.6)$$

Then Pareto optimality requires only that $\sum_{i \in C} U_i = V_{C,\mathcal{P}}$ and the game is equivalent to a TU game with $V_{C,\mathcal{P}}$ as the value of embedded coalition $C, \mathcal{P}$. Suppose moreover that $V_{C,\mathcal{P}} = V_{C,\mathcal{R}}$ for any two partitions $\mathcal{P}, \mathcal{R}$ with $C \in \mathcal{P}, C \in \mathcal{R}$. (That is, the value of the coalition is independent of the partition in which it is embedded.) Extending the terminology of McCain (2009, Chaps. 11–13), a game with these two properties is a *proper* game and is equivalent to a TU game in coalition function form. Note, however, that for such a game, the utility possibility set is only weakly convex, so that some mathematical operations for strictly convex sets may not apply to it.

However, we might instead adopt the view from the Paretian tradition in economics that no meaningful interpersonal comparison of utilities is possible. Then, even if side payments can occur, $\sum_{i \in C} U_i$ is not meaningful. The utility numbers are arbitrary indices of levels of preference, unique only

---

[3]Given $f_{\mathcal{P}}(\{U_i\}_{i \in N})$, we may characterize $f_{C,\mathcal{P}}(U_C)$ when U_j is known for $j \notin C$. If we follow the predominant tradition in cooperative game theory and associate the value of a coalition with the assurance value, then the U_j will correspond to the minimum value of $\max_{U_i} \sum_{i \in C} \lambda_i U_i$; i.e., in place of Eq. (3.3) we have $\max_{U_i} \min_{U_j} \sum_{i \in C} \lambda_i U_i$ subject to $f_{\mathcal{P}}(\{U_i\}_{i \in C}) \le 0, j \notin C$. If instead we apply Assumptions CR and CA, then $f_{C,\mathcal{P}}(\{U_i\}_{i \in C})$ is characterized at the Nash equilibrium of noncooperative play among the coalitions in partition $\mathcal{P}$.

up to a monotonic transformation. *Only* the weighted sum $\sum_{i \in C} \lambda_i U_i$ is meaningful, since it enforces an arbitrary interpersonal comparison in the form of the distributional weights. Let us call this the Paretian case. Even if one's principles are Paretian, the intermediate case gains interest because it is comparable to both the Paretian case and the proper case.

3.3.1. *Superior Games*

In the theory of TU games in coalition function form, it is often assumed that the game is superadditive; i.e., if two coalitions are merged, the value of the merged coalition is no less than the sum of the values of the two coalitions separately. This reflects the argument (von Neumann and Morgenstern, 2004, pp. 241–242) or intuition that any strategies and payoffs available to the two coalitions separately will be available also to the merged coalition. Since the value is not defined for NTU games, the corresponding property for NTU games cannot be defined in this way.

Let Γ be a NTU game in partition function form. Let $B \in \mathcal{P}$, $C \in \mathcal{P}$, $f_{B,\mathcal{P}}$ and $f_{C,\mathcal{P}}$ their respective constraint functions, and $A = B \cup C$, $A \in \mathcal{R}$, with $f_{A,\mathcal{R}}$ the corresponding constraint function. Here $\mathcal{R}$ is such that for any $S \in \mathcal{R}$, $S \neq A$, $S \neq B$. $S \in \mathcal{R}$.[4] It will be helpful to characterize a constraint function that will identify imputations simultaneously feasible for both B and C. Accordingly, let $f_{B,C,\mathcal{P}}(\{U_i\}_{i \in A \cup B}) = \max[f_{B,\mathcal{P}}(U_B), f_{C,\mathcal{P}}(U_C)]$. Then $f_{B,C,\mathcal{P}}(\{U_i\}) \leq 0$ will imply that $\{U_i\}$ is feasible for both B and C.

Definition 3.17. Suppose Γ is such that for every $A, B, C, \mathcal{P}$ and $\mathcal{R}$ as characterized in the previous paragraph, and for all $\{U_i\}, f_{B,C,\mathcal{P}}(\{U_i\}) \geq f_{A,\mathcal{R}}(\{U_i\})$. Then we will say that Γ is a *superior* game. If for all $\{U_i\}, f_{B,C,\mathcal{P}}(\{U_i\}) > f_{A,\mathcal{R}}(\{U_i\})$, we may say that Γ is strictly superior.

Remark. The property of superiority would seem to correspond to the plausible argument in the previous passage as superadditivity does for TU games.

3.3.2. *Example: A Common-Property Resource*

A group of n agents make use of a common-property resource. The agents might be countries sharing a fishery, local jurisdictions sharing the water

---

[4]That is, in the terminology of McCain 2009 (p. 172), $\mathcal{P}$ is particulate with respect to $\mathcal{R}$ and A.

Value Solutions in Cooperative Games

resources of a watershed, or some similar group. For agent i the quantity used is y_i and

$$Y = \sum_{i=1}^{N} y_i. \tag{3.7}$$

The net benefit to agent i is well approximated by the linear-quadratic function

$$u_i = \alpha_i y_i - \frac{\beta_i}{2} y_i^2 - \gamma_i Y. \tag{3.8}$$

Assumption. $\forall i \exists z_i \ni z_i > 0$ and

$$\beta_i = \frac{\alpha_i - \gamma_i}{z_i} > 0. \tag{3.9}$$

Since

$$\frac{\partial u_i}{\partial y_i} = \alpha_i - \gamma_i - \beta_i y_i \tag{3.10}$$

for any singleton coalition, the dominant strategy is

$$y_i = z_i = \frac{\alpha_i - \gamma_i}{\beta_i} \tag{3.11}$$

and thus, assuming the fine partition $\mathcal{F}$ and a Nash equilibrium among the singleton coalitions,

$$Y^{\mathcal{F}} = \sum_{i=1}^{N} \frac{\alpha_i}{\beta_i} - \sum_{i=1}^{N} \frac{\gamma_i}{\beta_i}. \tag{3.12}$$

Thus for the fine partition,

$$u_i^{\mathcal{F}} = \alpha_i \left(\frac{\alpha_i - \gamma_i}{\beta_i} \right) - \frac{1}{2\beta_i} (\alpha_i - \gamma_i)^2 - \gamma_i \sum_{j=1}^{N} \left(\frac{\alpha_j - \gamma_j}{\beta_j} \right). \tag{3.13}$$

Consider a partition $\mathcal{P} = \{C_1, C_2, \ldots, C_m\}, m < N$. Let m_k be the cardinality of C_k; thus $\sum_{k=1}^{m} m_k = N$. For k, let $Y_k = \sum_{\substack{j=1 \\ j \notin C}}^{N} y_j$. Thus

$$Y = Y_k + \sum_{j \in C_k} y_j. \tag{3.14}$$

Assuming noncooperative play among the coalitions, the utility possibility frontier for coalition k will be characterized by

$$\max \sum_{i \in C_k} \lambda_i [\alpha_i y_i - \beta_i y_i^2 - \gamma_i Y] \qquad (3.15a)$$

subject to

$$\sum_{i \in C_k} y_j \leq Y - Y_k \qquad (3.15b)$$

for some distributional weights λ_i. Thus, for this game,

$$u_i = \alpha_i y_i - \beta_i y_i^2 - \gamma_i \sum_{j=1}^{n} y_j \qquad (3.15c)$$

inverting,

$$y_i = g_i(u_i) \qquad (3.15d)$$

where g is a real, positive solution to the quadratic equation

$$-\beta_i y_i^2 + (\alpha_i - \gamma_i) y_i - u_i - \gamma_i \sum_{\substack{j=1 \\ j \notin C}}^{n} y_j = 0. \qquad (3.15e)$$

(If the solution is irrational then u_i is not feasible.) Then

$$f_{C_k, \mathcal{P}}(\{u_j\}) = \sum_{j \in C_k} g_i(u_i) + Y_k - Y \qquad (3.15f)$$

where Y_k is determined by the decisions of other coalitions. We see that, in principle, $f_{C_k, \mathcal{P}}$ is not independent of the decisions of others, but assuming Nash equilibrium play among the coalitions, we may identify Y_k with their predictable best responses.

Without loss of generality let

$$\sum_{i \in C_k} \lambda_i = \frac{m_k}{N} \quad \left(\text{thus} \sum_{i=1}^{N} \lambda_i = 1 \right). \qquad (3.16a)$$

Configuring the Lagrangean expression,

$$\mathcal{L} = \sum_{i \in C_k} \lambda_i [\alpha_i y_i - \beta_i y_i^2 - \gamma_i Y] + \mu_k \left[Y - Y_k - \sum_{j \in C_k} \right] y_j. \qquad (3.16b)$$

Among the necessary conditions from 3.15a are

$$\frac{\partial \mathcal{L}}{\partial y_i} = \lambda_i[\alpha_i - \beta_i y_i] - \mu_k \leq 0 \tag{3.16c}$$

$$\frac{\partial \mathcal{L}}{\partial Y} = -\sum_{i \in C_k} \lambda_i \gamma_i + \mu_k \leq 0. \tag{3.16d}$$

Remark. Here we assume an interior solution, i.e., $y_i > 0$ for all i. A corner solution would assign a rate of exploitation of zero to one or more agents. In the absence of side payments, it does not seem that this would be consistent with individual rationality. A formal demonstration will not be given here, however. Thus, assuming an interior solution and combining,

$$\alpha_i - \beta_i y_i - \sum_{j \in C_k} \frac{\lambda_j}{\lambda_i} \gamma_j = 0. \tag{3.16e}$$

Thus,

$$y_i - \frac{1}{\beta_i} \left(\alpha_j - \sum_{j \in C_k} \frac{\lambda_j}{\lambda_i} \gamma_j \right) \tag{3.16f}$$

Remark. In place of γ_i in Eqs. (3.10) and (3.11), we have a weighted sum of all parameters γ_j for all members of the coalition C_k. To characterize the utility possibility frontier for C_k, allow the distributional weights to vary over all values consistent with Eq. (3.16f) and a non-negative value of y_i. Noting that the weight on γ_i, $\frac{\lambda_i}{\lambda_i} = 1$, we may rewrite Eq. (3.16f) as

$$y_i = \frac{1}{\beta_i} \left(\alpha_i - \gamma_i - \sum_{\substack{j \in C_k \\ j \neq i}} \frac{\lambda_j}{\lambda_i} \gamma_j \right) \tag{3.16g}$$

so we see that each agent in the coalition will reduce his use of the common property resource below the dominant strategy level by an amount that depends on the values of the parameter γ for other members of the coalition and on the relative distributional weights. From Eq. (3.16f), we have

$$Y = \sum_{i=1}^{N} \frac{1}{\beta_i}(\alpha_i - \gamma_i) - \sum_{C_k \in \mathcal{P}} \sum_{j \in C_k} \frac{h_j}{\beta_j} \tag{3.16h}$$

where

$$h_j = \sum_{\substack{s \in C_j \\ s \neq j}} \frac{\lambda_s}{\lambda_j} \gamma_s \tag{3.17a}$$

and

$$C_i = C \in \mathcal{P} \ni i \in \mathcal{C}. \tag{3.17b}$$

Benefits for agent i then are

$$u_i^{\mathcal{P}} = \frac{\alpha_i}{\beta_i}(\alpha_i - \gamma_i - h_i) - \frac{1}{2\beta_i}(\alpha_i - \gamma_i - h_i)^2$$

$$- \gamma_i \left[\sum_{j=1}^{N} \frac{1}{\beta_j}(\alpha_j - \gamma_j) - \sum_{C_k \mathcal{P}} \sum_{j \in C_k} \frac{h_j}{\beta_j} \right]$$

$$= u_i^{\mathcal{F}} + \gamma_i \sum_{j=1}^{N} \frac{1}{\beta_j} \sum_{\substack{s \in C_j \\ s \neq i \\ s \neq j}} \frac{\lambda_s}{\lambda_j} \gamma_s - \frac{\gamma_i h_i}{\beta_i} - \frac{h_i^2}{2\beta_i}. \tag{3.16k}$$

Lemma 3.7. *The game defined in this way is superior.*

Proof. From Eq. (3.17b), the Nash equilibrium among the coalitions in this game can be characterized as

$$\max \sum_{C_k \in \mathcal{P}} \sum_{i \in C_k} \lambda_i [\alpha_i y_i - \beta_i y_i^2 - \gamma_i Y] \tag{3.18a}$$

or equivalently

$$\max \sum_{i=1}^{N} \lambda_i [\alpha_i y_i - \beta_i y_i^2 - \gamma_i Y] \tag{3.18b}$$

subject to

$$\sum_{j \in C_k} y_j \leq Y - Y_k \forall C_k \in \mathcal{P} \tag{3.18c}$$

and

$$\sum_{i \in C_k} \lambda_i = \frac{m_k}{N} \forall C_k \in \mathcal{P}. \tag{3.18d}$$

Now consider partition $\mathcal{R} = \{B_1, B_2, \ldots, B_{m+1}\} \ni B_r \cup B_{r+1} = C_r \in \mathcal{P}$, for $q < r, B_q = C_q$, and for $q > r + 1, B_q = C_{q+1}$. That is, $\mathcal{P}$ is formed from $\mathcal{R}$ by the merger of two coalitions. To characterize the Nash equilibrium for $\mathcal{R}$ we will again maximize (3.15a), but for the constraint $\sum_{j \in C_r} y_j \leq Y - Y_r$ we substitute two constraints $\sum_{j \in B_r} y_j \leq Y - Y_r^*$ and $\sum_{j \in B_{r+1}} y_j \leq Y - Y_{r+1}^*$, with Y_r^* and Y_{r+1}^* appropriately defined. We then find that any vector of y_i that satisfies these constraints will satisfy constraints in Eq. (3.15b). Similarly, for $\sum_{i \in C_r} \lambda_i = \frac{m_r}{N}$ we may substitute the two $\sum_{i \in B_r} \lambda_i = \frac{m_r^*}{N}$ and $\sum_{i \in B_{r+1}} \lambda_i = \frac{m_{r+1}^*}{N}$, where m_r^* and m_{r+1}^* are the cardinalities of B_r and B_{r+1}, respectively. Again, we find that any vector of distributional weights λ_i that satisfies the revised constraints will satisfy Eq. (3.15b). In short, any point in the utility possibility frontier for B_r and B_{r+1} will be within or at the utility possibility frontier for C_r, and since that is true in general the game is superior.   □

3.3.3.   *Example: A Cartel with Differentiated Products*

Suppose that a group of n firms produce differentiated products with

$$Q_i = \alpha_i \prod_{j=1}^{n} p_j^{\beta_{j,i}} \qquad (3.19)$$

where Q_i is the quantity demanded of the product of firm i and $\alpha_i > 0, \beta_{i,i} < -1$, and $\beta_{j,i} > 0$ for $j \neq i$; moreover,

$$C_i = c_i Q_i + F_i \qquad (3.20)$$

where C_i is the total cost for firm i, c_i is a constant unit cost and F_i a fixed cost. For the fine partition each firm competes as a singleton coalition and the price rule

$$p_i^* = \frac{\beta_{i,i} c_i}{1 + \beta_{i,i}} \qquad (3.21)$$

is a dominant strategy. This dominant strategy solution will be recognized as an application of the "Lerner Rule" (Lerner, 1934) and is a consequence of the assumption of constant elasticities; it will apply for any singleton coalition regardless of the coalition structure of the remaining firms.

Now suppose $S = \{1, 2, 3, \ldots, m\}$ form a cartel. For a TU model, the cartel's prices would be such as to maximize the cartel's profits, and the profits might be redistributed among the participants via side payments.

Suppose instead that side payments cannot occur, and each firm simply keeps the profits that it accrues. Then it may be necessary for the cartel to tolerate individual firm prices that deviate from those that would maximize the cartel's total profit; however, the cartel's utility possibility frontier will be a hypersurface of vectors $(\pi_1, \pi_2, \ldots, \pi_m)$ with each point generated by a vector $(p_1, p_2, \ldots, p_m)$, where π_i is the profit and p_i is the price for firm i.

For this example U_i can be identified with firm profits, $(p_i - c_i)Q_i - F_i$. Since (3.21) is a constraint that applies for each firm separately, the constraint function for the cartel will be

$$f_{C,\mathcal{P}}(\{U_i\}_{i=1}^m) = \max_i \left(Q_i - \alpha_i \prod_{j=1}^n p_j^{\beta_{j,i}} \right). \tag{3.22}$$

To characterize the cartel's utility-possibility frontier, we consider the Lagrangean function,

$$\mathcal{L} = \sum_{i=1}^m \lambda_i [(p_i - c_i)Q_i - F_i] + \sum_{i=1}^m \mu_i \left(Q_i - \prod_{j=1}^n p_j^{\beta_{j,i}} \right). \tag{3.23}$$

From the necessary conditions for a maximum we obtain

$$p_i = \frac{\beta_{i,i} c_i}{1 + \beta_{i,i}} - \sum_{\substack{j=1 \\ j \neq i}}^m \frac{\lambda_j}{\lambda_i} \frac{\beta_{i,j}}{1 + \beta_{i,i}} (p_j - c_j). \tag{3.24}$$

This generalizes the "Lerner rule" and is a dominant strategy for the cartel in its noncooperative play against nonmembers. The solution to this system, for given $\frac{\lambda_j}{\lambda_i}$, determines the cartel prices for the various firms, and thus, via (3.19), the corresponding outputs and profits (utilities). Letting the distributional weights λ_j vary over the m-dimensional simplex, we characterize the utility possibility frontier for the cartel. Note that, applying the generalized Le Chatelier principle to (3.22), we may immediately note that this is a superior game.

3.4. Chapter Summary

This chapter has discussed the representation of games in partition function form, which can allow the modeling of externalities explicitly in the values of coalitions. For this purpose, we allow for noncooperative interactions among coalitions, although relations within the coalitions are supposed to be cooperative. While there is a long tradition of discussion of TU games in these terms, some care needs to be taken in the definition of concepts such as symmetry, efficiency and superadditivity in this context. The presence of externalities, and particularly of negative externalities, introduces complexity into the relationships among those properties. For solution concepts such as the core and the nucleolus, which use the stability of an imputation against deviations, it is necessary in some way to specify the partition that follows from a deviation, expressed as a successor function. The discussion of NTU games in partition function form seems to be novel. Borrowing some ideas from welfare economics, we characterize a utility possibility frontier for each embedded coalition, and explore two applications: a cartel with differentiated products and a group of users of a common-property resource.

Chapter 4

A Shapley Value Algorithm for Games in Partition Function Form

Of all value solutions for transferable utility (TU) games in coalition function form, the most widely studied surely is the Shapley value. Shapley gives two interpretations of the value imputation: first, as a combinatorial formula (which uniquely possesses some attractive mathematical properties), and second, as an expected value of the marginal contribution of the agent to a coalition. The major content of Shapley's classic paper is a proof that only values calculated according to that formula will satisfy certain plausible conditions. For the second interpretation, suppose that the agents in the game are consulted in a particular order. Coalition C_i is the coalition of the first i such agents. The agent indexed at i is awarded his "marginal contribution," the value of C_i minus the value of $C_i \backslash \{i\}$. These awards will be different if the agents are taken in a permuted order. Let all permutations be equally probable. Then the Shapley value for an individual is the expected value of the marginal contribution over all permutations.

The Shapley value can be interpreted, in computational terms, as an algorithm. This chapter explores an algorithm for a value solution for TU games in partition function form that computes a quasi-Shapley value based on one particular property of the Shapley value, marginal valuation.

4.1. Shapley's "Algorithm"

We might construe either of these interpretations as a computational algorithm. Any combinatorial formula is a (very simple instance of a) computational algorithm, and the expected value interpretation can also be understood as an instance of a computational algorithm. For this chapter, we will reverse the figure and the ground: that we might say instead that the solution to a cooperative game is a computational algorithm. A more complex game could call for a more complex algorithm and the algorithm might produce more complex products than simply an imputation. In this chapter, we will be concerned with a more complex algorithm reducible to the second, expected value interpretation of the Shapley value. (We may note in passing that for more complex games the two interpretations may not coincide as they do in Shapley's discussion.)

Here is an algorithmic restatement of the marginal value interpretation of the Shapley value. First let the players in the game be indexed as $i = 1, 2, \ldots, N$, and let $m = f_j(i)$ be a permutation of the order of the indices. Let the value of a set S of players be $v(S)$. Let the set of all distinct permutations be indexed as $j = 1, 2, \ldots, K$. Now execute

Loop for $j = 1, \ldots, K$.
　　Loop for $m = 1, \ldots, N$

$$i = f_j^{-1}(m), \quad s(j,i) = v\left(\{\iota\}_{f_j^{-1}(t)}^{-1} = 1^i\right) - v\left(\{\iota\}_{f_j^{-1}(t)=1}^{i-1}\right).$$

Remark. $s(j,i)$ will be called the partial value of i for j and may be thought of as the marginal contribution of i to the coalition $\{\iota\}_{f_j^{-1}(t)=1}^{i-1}$.

　　End loop
　End loop
　Loop for $i = 1, \ldots, N$
　　$s(i) = \text{Averate}_{j=1}^{K} \, s(j,i)$
　End loop

Remark. $s(i)$ is the Shapley value for Player i.

4.2.　Maskin's Proposal

An important recent contribution was by Maskin (2004). In the Shapley value, the imputed payment to an individual is derived from the agent's marginal contributions. This property is called "marginality." As Maskin notes, for a superadditive game in coalition function form, $v(\{\iota\}_{f_j^{-1}(t)=1}^{i}) - v(\{\iota\}_{f_j^{-1}(t)=1}^{i-1}) \geq v(\{i\})$, so there is no reason for Player i to refuse to join the coalition $\{\iota\}_{f_j^{-1}(t)-1}^{i-1}$. However, Maskin considers games in partition function form. In such a game, if i benefits from a positive externality created by $\{\iota\}_{f_j^{-1}(t)=1}^{i-1}$ (e.g., if $\{\iota\}_{f_j^{-1}(t)=1}^{i-1}$ is a coalition to produce a public good) i might be better off to decline to join, instead remaining as a singleton free rider or joining with other agents ι with $f_j(\iota) > m$ as a separate coalition. There are a number of proposals for extension of the Shapley value to games in partition function form, some of which weaken or lack the property of marginality. In Maskin's paper, the imputations lack marginality and substitute a concept related to opportunity cost. Maskin suggests that marginality attributes overwhelming bargaining power to the new recruit to a coalition, and he instead attributes overwhelming bargaining power to an existing coalition. But Maskin's solution differs from Shapley's in another important way: he does not assume that the grand coalition will form, and allows agents, as they are consulted, to join or refuse to join an existing coalition. Thus, the algorithm generates a prediction of the partition that will form as well as an imputation of payments to the agents. (In fact the prediction of the partition formed is not independent of the permuted order in which the agents are consulted, so that Maskin's prediction is a fairly complex object.) As with Shapley, Maskin's solution can be construed as a computational algorithm, though he does not explicitly present it as such.

Another possibility, not addressed by Maskin, is that if the game is not superadditive, it is possible that $v\big(\{\iota\}_{f_j^{-1}(t)=1}^{i}\big) - v\big(\{\iota\}_{f_j^{-1}(t)=1}^{i-1}\big) < v(\{i\})$, so that i might be better off remaining as a singleton. Thus, in either case, the grand coalition may not form, and Maskin argues this will occur in games with positive externalities in particular.

A difficulty is that, for games in partition function form, this sequential addition of agents to *the partition formed by agents previously added* may result in the formation of different partitions for different permutations. Accordingly, Maskin's approach actually gives rise to a large family of solutions, with one solution for each j and i in the double-loop algorithm. More generally, multiplicity of solutions seems to be a consequence of introducing externalities into the analysis (de Clippel and Serranno 2008a, p. 1414; McQuillin, 2008, p. 698). Moreover, Maskin's results do not extend to games of more than three players (de Clippel and Serranno, 2008b, p. 1414). This illustrates the intrinsic complexity of the problem, since some real world applications are likely to involve many more than three players.

The research reported in this chapter, however, will draw on a slightly different interpretation of Shapley values. On one hand, marginal valuation will be retained. On the other hand, decisions to enter an existing coalition or to remain a singleton will be made by the individuals, in the permuted order, to maximize their marginal valuation in a subgame perfect sequence of decisions.

4.3. An Extended Shapley Algorithm

This chapter explores a solution to a game in partition function form, an algorithm that differs from Maskin's in that it retains marginality and conceives a solution somewhat differently. The solution is based on the following ideas:

  (I) It computes a probability distribution over the partitions that may be formed.
 (II) For each partition with a positive probability, it imputes a value for each agent. The imputation is conditional on the formation of that partition.
(III) The value is marginal, that is, it is the expected value of the agent's marginal contribution to the coalition of which he is a member, with the expected value computed over all permutations of the agents that yield the partition, with the permutations considered as equiprobable. (This is the sense in which it is a quasi-Shapley solution.)
 (IV) The total of the values of agents in a coalition (imputed for a particular partition) just exhausts the value of the coalition in the partition.
  (V) At each step for a given permutation, the agent may choose to join any coalition already formed, or to enter the game as a singleton coalition. (This is suggested by Maskin's model.)
 (VI) When these decisions are considered as a game in extensive form, the decisions are subgame perfect best responses.

As Shapley notes, the average of the partial values can be interpreted as the expected value of the agent's marginal contribution, if the players are added in some permuted order and permutations are all equally probable. For this chapter, the solution to a game will be probabilistic and will be based on the treatment of all permutations as equally probable. With that assumption, the solution will comprise (1) a list of partitions with the probabilities that those partitions will be formed, and (2) associated with each partition that has a positive probability, a set of payments to each player, constituting the player's expected value of payoffs over all permutations that result in the formation of that partition; that is, a value solution for the game conditional on that partition.

The double-loop algorithm above will be modified as follows. The outer loop is unmodified. In the inner loop, at each step a permutation j is given. For $m = 1$, agent $f_j^{-1}(m)$ can only form a singleton coalition. For $m > 1$, agent $f_j^{-1}(m)$ may decide to remain a singleton, so that at least two distinct partitions may arise. Agent $f_j^{-1}(m)$ will therefore confront a partition that is the consequence of choices made by other agents with permuted indices $f(\iota) < m$. Denote that partition as $P_{j,m}$ We suppose that $f_j^{-1}(m)$ can decide to join any of those existing coalitions, or remain a singleton, so that a still larger family of partitions will be confronted by the next agent $f_j^{-1}(m + 1)$. This sequence of decisions is treated as a game in extensive form, and for each such sequence a set of partial values is calculated for each terminus of the game, and the partial values are the payoffs for the game in extensive form. The partial values are computed as marginal contributions, with coalitions that are elements of $\mathcal{P}_{j,N}$ valued as embedded in $\mathcal{P}_{j,N}$, while others are valued as embedded in $\mathcal{P}_{j,m+1}$. If there is an unique subgame perfect equilibrium of the extensive game determined in this way, $\mathcal{P}_{j,N}$ is the partition predicted for that permutation. If there are two or more subgame perfect equilibria, they are considered equiprobable. For the solution of a particular game, the probability of a given partition is the sum of the probabilities of permutations for which it is the predicted outcome, or of fractional probabilities for permutations that do not have unique predictions. For each partition, the values for individual agents are the expected values of their partial values for permutations in which that partition is formed, conditional on the formation of that partition.

The Shapley value has two properties that are crucial for this study. The first is marginality. The second one, which may seem trivial, is that the values exactly exhaust the value of the grand coalition. The second of these is complicated for games in partition function form, however, if the grand coalition does not form. Certainly the total value for the individuals cannot exceed the total value of the coalitions that form: this follows from the definition of an imputation. Following Aumann and Dreze (1974), we might further require that each tub sits on its own bottom, that is, that the values of the members of a coalition exactly exhaust *the value of the coalition of which they are members*. Much of the modest literature on extending the Shapley value has chosen the first of those assumptions, but this research instead follows Aumann and Dreze.

Computer code for the application of this algorithm is given available from the author on his website, http://faculty.lebow.drexel.edu/mccainr/top/eco/wps/program.pdf.

4.4. Some Examples

The above discussion will be made more clear by examples. This section discusses some small-scale games, with computed solutions and some interpretation in terms of subgame perfect decisions to join the incremental coalitions.

4.4.1. *An Illustrative Example*

We first consider a simplified example of a public-goods production game, Game 4.1, NIMBY. NIMBY is a three-person game in which one of the three agents may produce a single unit of a public good that benefits all three, but results in a net loss to the producer. Thus, the noncooperative equilibrium would (as usual) correspond to zero production of the public good, i.e., "not in my back yard." NIMBY is shown in partition function form in Table 4.1.

NIMBY provides an example of dominance cycles. For any imputation, the grand coalition will be dominated by one of the two-versus-one coalitions, since any agent can get a payoff of 6 by seceding from the grand coalition and taking advantage of the positive externality from the other two. However, each two-versus-one partition is dominated by the grand coalition with a payoff allocation that gives more than 8 (total) to the members of the two-person coalition and more than 6 to the singleton. The quasi-Shapley solution, given as Table 4.2, indicates that these peculiarities may be no obstacle to the formation of the grand coalition if coalitions are formed as in the quasi-Shapley algorithm.

Table 4.1. Game 4.1. NIMBY.

	Coalitions	Coalition values		
1	$\{a\}, \{b\}, \{c\}$	3,	3,	3
2	$\{a, b\}, \{c\}$	8,	6	
3	$\{a, c\}, \{b\}$	8,	6	
4	$\{b, c\}, \{a\}$	8,	6	
5	$\{a, b, c\}$	16		

Table 4.2.   The solution to Game 4.1.

For partition 1 the probability weight is 0
For partition 2 the probability weight is 0
For partition 3 the probability weight is 0
For partition 4 the probability weight is 0
For partition 5 the probability weight is 1
The conditional expected value payoffs are
    For agent a: 5.333333
    For agent b: 5.333333
    For agent c: 5.333333

Table 4.3.   Game 4.2: A strong positive externality game.

1	$\{a\}, \{b\}, \{c\}, \{d\}$	2, 2, 2, 3
2	$\{a, b\}, \{c\}, \{d\}$	10, 3, 4
3	$\{a, c\}, \{b\}, \{d\}$	10, 3, 4
4	$\{a\}, \{b, c\}, \{d\}$	3, 10, 4
5	$\{a, b, c\}, \{d\}$	15, 14
6	$\{a, d\}, \{b\}, \{c\}$	12, 4, 4
7	$\{a\}, \{b, d\}, \{c\}$	3, 10, 4
8	$\{a\}, \{b\}, \{c, d\}$	3, 3, 10
9	$\{a, b, d\}, \{c\}$	15, 14
10	$\{a, b\}, \{c, d\}$	12, 12
11	$\{a, c, d\}, \{b\}$	15, 14
12	$\{a, c\}\{b, d\}$	12, 12
13	$\{a, d\}\{b, c\}$	14, 14
14	$\{a\}\{b, c, d\}$	14, 15
15	$\{a, b, c, d\}$	30

4.4.2.   *A Game with Strong Positive Externalities*

Next consider Game 4.2, a superadditive game with some strong positive externalities. Game 4.2 is displayed in partition function form in Table 4.3. The computed solution for this game is shown in full in Table 4.4.

We see that in this game the formation of the efficient grand coalition is quite probable, with a probability of $\frac{5}{6}$, but that there is a $\frac{1}{6}$ probability that the less efficient partition 13, $\{\{a, d\}, \{b, c\}\}$ will form, and moreover, in that case agents b and c will do somewhat better. The probabilities reflect the fact that in 20 of the 24 permutations of the six agents, a sequence of choices leading to the grand coalition is subgame perfect, but in four others, the subgame perfect sequence leads to partition 13. One such permutation is $\{d, a, b, c\}$. At the first stage agent d forms a singleton coalition in the

Table 4.4.   Solution to the strong positive externality game.

For partition 1 the probability weight is 0
For partition 2 the probability weight is 0
For partition 3 the probability weight is 0
For partition 4 the probability weight is 0
For partition 5 the probability weight is 0
For partition 6 the probability weight is 0
For partition 7 the probability weight is 0
For partition 8 the probability weight is 0
For partition 9 the probability weight is 0
For partition 10 the probability weight is 0
For partition 11 the probability weight is 0
For partition 12 the probability weight is 0
For partition 13 the probability weight is 0.1666667
The conditional expected value payoffs are
    For agent a: 6.5
    For agent b: 7
    For agent c: 7
    For agent d: 7.5
For partition 14 the probability weight is 0
For partition 15 the probability weight is 0.8333333
The conditional expected value payoffs are
    For agent a: 8
    For agent b: 6.9
    For agent c: 6.9
    For agent d: 8.2

---

fine partition $\{\{a\}, \{b\}, \{c\}, \{d\}\}$. Conventionally we think of this as adding agent d to a null partition so that the marginal value is the value of $\{d\}$ minus the zero value of the null coalition, that is, 3. At the second stage agent a has the choice of remaining as a singleton, which has the value of 2 in the fine partition, or of joining with agent d to form $\{\{a, d\}, \{b\}, \{c\}\}$, partition 6. The value of $\{a, d\}$ in partition 6 is 12, but agent a can rationally anticipate that if he forms $\{a, d\}$, agents b and c will form coalition $\{b, c\}$, and in partition $\{\{a, d\}, \{b, c\}\}$, partition 13, Coalition $\{a, d\}$ is valued at 14. Accordingly, in the subgame perfect equilibrium for this permutation, the marginal value of agent a is $14 - 3 = 11$, and agent a forms the two-person coalition. At the third stage agent b has the choice of joining the two-person coalition to form $\{\{a, b, d\}, \{c\}\}$, partition 9, or of remaining as a singleton in partition 6. Coalition $\{a, b, d\}$ is valued at 15 in partition 9, so that the marginal value of agent b in that case is $15 - 14 = 1$. On the other hand, the value of the singleton is 4 (thanks in part to the positive externality from

$\{a, d\}$) and the singleton is agent 2's subgame perfect choice. Finally, in step 4, agent c can choose to join $\{a, d\}$ to form $\{\{a, c, d\}, \{b\}\}$, partition 11, to join $\{b\}$ to form $\{\{a, d\}, \{b, c\}\}$, partition 13, or to remain as a singleton in partition 6. The marginal values are respectively $15 - 14 = 1$, $14 - 4 = 10$, or 4. Clearly the second of these is the best response.

However, the inefficiency in this case is not strictly speaking an instance of free-rider or holdout behavior. Instead, it occurs because the algorithm allows no opportunity for coalitions $\{a, d\}$ and $\{b, c\}$ to merge. Such a merger would be inconsistent with marginality, since there is no clear *marginal* imputation to assign the benefits of the merger to the individuals in the grand coalition. This could be offered as an argument *against* a marginal imputation scheme for games in partition function form. However, in an individualistic society, coalitions often are recruited step-by-step, one member at a time, and the marginal imputation scheme captures this.

Luce and Raiffa (1957) interpret the Shapley value as an arbitration scheme, along the following lines: the arbitrator, not himself a player in the game, might say to the agents in the game "If you sign a contract to form the grand coalition with your Shapley values, then you receive those values with certainty. If you decline the contract and rely on individualistic recruitment, then the Shapley values are your expected value, but it is a risky expectation. You have, on the average, nothing to lose by signing the contract now, and if you are at all risk averse, you have something to gain." In this case, however, agents b and c can respond "but with individualistic recruitment, there is a one-sixth probability that partition 13 will form, and we will be paid 7 rather than 6.9, so that our expected value payouts from the individualistic recruitment will be 6.9167, and since we are risk neutral we will not sign." Even in a superadditive game, some coalitions may benefit from an inefficient arrangement, and this means that some inefficiency may be persistent.

4.4.3. *A Game with Mild Positive Externalities*

For contrast with Games 4.1 and 4.2, consider Game 4.3, which is also superadditive and has positive externalities, but the externalities are relatively slight. Game 4.3 is shown as Table 4.5. Otherwise it is very similar to Game 4.2. The computed solution is shown as Table 4.6, showing only the partition with a positive probability. The solution attributes probability 1 to the efficient grand coalition.

Table 4.5.   Game 4.3.

1	$\{a\},\{b\},\{c\},\{d\}$	2, 2, 2, 3
2	$\{a,b\},\{c\},\{d\}$	10, 3, 3
3	$\{a,c\},\{b\},\{d\}$	10, 3, 3
4	$\{a\},\{b,c\},\{d\}$	3, 10, 4
5	$\{a,b,c\},\{d\}$	20, 5
6	$\{a,d\},\{b\},\{c\}$	12, 4, 4
7	$\{a\},\{b,d\},\{c\}$	3, 10, 3
8	$\{a\},\{b\},\{c,d\}$	3, 3, 10
9	$\{a,b,d\},\{c\}$	20, 5
10	$\{a,b\},\{c,d\}$	12, 12
11	$\{a,c,d\},\{b\}$	22, 6
12	$\{a,c\}\{b,d\}$	12, 12
13	$\{a,d\}\{b,c\}$	14, 14
14	$\{a\}\{b,c,d\}$	5, 20
15	$\{a,b,c,d\}$	30

Table 4.6.   The solution of Game 4.3.

For partition 15 the probability weight is 1
The conditional expected value payoffs are
    For agent a: 7.75
    For agent b: 6.75
    For agent c: 7.416667
    For agent d: 8.083333

We see that, in this case, the positive externalities are not an obstacle to the formation of the efficient grand coalition. The mild asymmetries in the coalition values result in some asymmetry in the expected value imputations. A comparison of this game with Game 4.2 illustrates the fact that, while the solution is sensitive to positive externalities, they will result in inefficiencies only if they are relatively pronounced.

4.4.4.   *A Game that is not Superadditive*

Consider Game 4.4, which is shown in Table 4.7. This game displays negative externalities and is not superadditive. Indeed, the only efficient partition in this game is the fine partition, $\{\{a\},\{b\},\{c\},\{d\}\}$. Game 4.4 is a four-person game suggested by Galbraith's (1952) hypothesis of countervailing power. Generally, formation of coalitions decreases the average payoff, but with the following exceptions: if either a and b or c and d form two-person coalitions, the members of the coalition gain a substantial

Table 4.7.   Game 4.4, A Non-superadditive Game.

1	$\{a\}, \{b\}, \{c\}, \{d\}$	5, 5, 5, 5
2	$\{a, b\}, \{c\}, \{d\}$	15, 1, 1
3	$\{a, c\}, \{b\}, \{d\}$	9, 5, 5
4	$\{a\}, \{b, c\}, \{d\}$	5, 9, 5
5	$\{a, b, c\}, \{d\}$	12, 5
6	$\{a, d\}, \{b\}, \{c\}$	9, 5, 5
7	$\{a\}, \{b, d\}, \{c\}$	5, 9, 5,
8	$\{a\}, \{b\}, \{c, d\}$	1, 1, 15
9	$\{a, b, d\}, \{c\}$	12, 5
10	$\{a, b\}, \{c, d\}$	9, 9
11	$\{a, c, d\}, \{b\}$	12, 5
12	$\{a, c\}\{b, d\}$	8, 8
13	$\{a, d\}\{b, c\}$	8, 8
14	$\{a\}\{b, c, d\}$	5, 12
15	$\{a, b, c, d\}$	16

advantage while imposing a large negative externality on the other, unorganized pair; while if both $\{a, b\}$ and $\{c, d\}$ coalesce, the outcome approximates the efficient fine partition but with some loss of its efficiency. This implies a complicated dominance cycle as the fine partition is unconditionally dominated (via $\{a, b\}$) by partition 2 and (via $\{c, d\}$ by partition 8. Both of these, in turn, are unconditionally dominated by partition 10, which in turn is dominated by the fine partition, although the fine partition is not accessible at one step from partition 10.

But the solution to this game assigns 0 probability to partition 1, and a 50% probability to partition 10, the countervailing power partition; with probabilities of $\frac{1}{8}$ each to partitions 3, 4, 6, and 7. The solution is given Table 4.6, showing only partitions with positive probabilities.

Again, why does this occur? Consider the natural permutation (a, b, c, d). At the first stage, agent a is added to the null coalition in partition 1, the fine partition $\{\{a\}, \{b\}, \{c\}, \{d\}\}$. The marginal value at this stage is the value of the singleton in the fine partition, 5. At the second step, agent b can join with agent a to form $\{a, b\}$ or can remain separate as a singleton. Suppose $\{a, b\}$ is formed. The marginal value attributable to agent b then is the value of $\{a, b\}$ minus the value of $\{a\}$, but again this is ambiguous. If $\{a, b\}$ is formed, then at the third step agent c will choose to remain a singleton (for agent c's marginal contribution to $\{a, b, c\}$ is negative, 12–15) and at the fourth stage agent d will join with agent c to form partition 10, $\{\{a, b\}, \{c, d\}\}$. Then, is the appropriate value of $\{a, b\}$ above construed to be $v(\{a, b\}, 2) = 15$, as partition 2 is formed at the

Table 4.8.   The solution to Game 4.4.

For partition 3 the probability weight is 0.125
The conditional expected value payoffs are
    For agent 1: 4.5
    For agent 2: 5
    For agent 3: 4.5
    For agent 4: 5
For partition 4 the probability weight is 0.125
The conditional expected value payoffs are
    For agent 1: 5
    For agent 2: 4.5
    For agent 3: 4.5
    For agent 4: 5
For partition 6 the probability weight is 0.125
The conditional expected value payoffs are
    For agent 1: 4.5
    For agent 2: 5
    For agent 3: 5
    For agent 4: 4.5
For partition 7 the probability weight is 0.125
The conditional expected value payoffs are
    For agent 1: 5
    For agent 2: 4.5
    For agent 3: 5
    For agent 4: 4.5
For partition 10 the probability weight is 0.5
The conditional expected value payoffs are
    For agent 1: 4.5
    For agent 2: 4.5
    For agent 3: 4.5
    For agent 4: 4.5

---

second stage, or $v(\{a,b\}, 10) = 9$? (The difference is the negative externality to $\{a,b\}$ from the formation of $\{c,d\}$.) Suppose we compute the marginal contribution of agent b as $v(\{a,b\}, 2) - v(\{a\}, 1) = 15 - 5 = 10$. Then, for this permutation, the total imputation to $\{a,b\}$ is 15, although in the partition actually formed $\{a,b\}$ produces only 9; and the total of all imputations is 24, though only 18 is produced in partition 10. Imputing values in this way would simply ignore the negative externality in the game. Evidently, we cannot satisfy either the "every tub on its own bottom" requirement or the less limiting requirement that the imputations cannot exceed total value unless we attribute the marginal contribution of agent 2 as $v(\{a,b\}, 10) - v(\{a\}, 1) = 9 - 5 = 4$. But this is also sound rational action theory, as 2 will make his decision on the rational expectation that coalition $\{c,d\}$ will be formed.

If we were to adopt a normative solution for this game, it might take the form of a social contract of individualism, i.e., for restraints that would prevent the formation of any coalitions. But the "game" of negotiating that social contract would then be a superior nontransferable utility (NTU) game in coalition function form, and a four-person bargaining problem. Both cooperative game models are appropriate to their respective purposes. It should be added, though, that Game 4.4 is something of a pure case for individualism, in that coalitions gain advantage *only* at the expense of others, through inefficient negative externalities. In the actual world, such externalities will often be incidental to coalitional production activities that are potential Pareto improvements. It is hardly likely that a social contract against coalitions would be a reasonable proposition in a more realistic context. However, the example may illustrate why countervailing power coalitions seem fairly common, as they are a reasonable response to negative externalities.

4.4.5.   *The Enigma Game*

Recall, from Chapter 3, Game 3.3, shown in Table 3.3. Although this game is superadditive, the fine partition is Pareto-efficient and Pareto-superior to any imputation in the grand coalition that approximates equal distribution, and the grand coalition is in turn Pareto-superior to any other partition. In this game, the total value of the grand coalition is less than that of the fine partition. This occurs because of the strong negative externalities in the game. These remarkable properties justify the reference to this as the Enigma game.

The quasi-Shapley solution for this game is shown in Table 4.9.

The solution in this case is as unremarkable as the game is enigmatic. Superadditivity leads to the formation of the grand coalition in the Enigma Game, with an equal distribution of the value, even though the fine partition

Table 4.9.   Quasi-Shapley solution to the enigma game.

For partition 15 the probability weight is 1
The conditional expected value payoffs are
    For agent 1: 1.75
    For agent 2: 1.75
    For agent 3: 1.75
    For agent 4: 1.75

dominates and is Pareto-superior to this solution. Here, again, we see that negative externalities can have powerful implications, even for superadditive games.

4.4.6. *An Exceptional Public Goods Game*

Game 4.5 is a five-person game of public goods production. For this game, each person can produce a varying quantity of a public good. If he produces none of the public good the individual's payoff is 50 plus the total quantity of public goods produced. If the individual produces the public good, he gives up a portion of the private good, G, so that his payoff is 50-G plus the total quantity of public goods produced. The public good is produced according to a Cobb–Douglas production function $Q = \alpha G^\beta$ with $\alpha = 3$ and $\beta = 0.75$. Notice that diminishing returns apply to the resources committed by an individual agent, not to the total resources committed by a coalition. With this specification, any coalition, even a singleton, will produce some of the public good; however only the grand coalition will produce an efficient quantity of the public good. The payoffs for different *forms* of partitions are shown in Table 4.10. Payoffs are shown *per member* in this case. We see that this game is exceptional in that there is no benefit for free rider behavior: the benefits of the public good are so great that a singleton holdout can increase his payoff by joining an existing coalition to produce public goods.

The computed solution (listing only partitions with positive probability) is shown as Table 4.11. In symmetrical games, the payoffs to players within a particular coalition are equal. Game 4.5 has five solutions, in each of which a four-person coalition is formed and the remaining agent remains as a singleton. Symmetrical payoffs for all agents are not possible because of the every-tub-on-its-own-bottom constraint, but as we see, payoffs are symmetrical within that constraint.

Table 4.10.   Payoffs per member in forms of partitions in Game 4.5.

1	$\{ijklm\}$	846.46			
2	$\{ijkl\}\{m\}$	306.96,	106.36		
3	$\{ijk\}\{lm\}$	107.17,	86.54		
4	$\{ijk\}\{l\}\{m\}$	69.75,	57.64	57.64	
5	$\{ij\}\{kl\}\{m\}$	58.33,	58.33,	54.64,	
6	$\{ij\}\{k\}\{l\}\{m\}$	55.17,	52.86,	52.86,	52.86
7	$\{i\}\{j\}\{k\}\{l\}\{m\}$	51.28,	51.28,	51.28,	51.28,	51.28

Table 4.11.   The solution of Game 4.5.

For partition 15 the probability weight is 0.2
The conditional expected value payoffs are
    For agent 1: 108.4875
    For agent 2: 108.4875
    For agent 3: 108.4875
    For agent 4: 108.4875
    For agent 5: 67.08
For partition 29 the probability weight is 0.2
The conditional expected value payoffs are
    For agent 1: 108.4875
    For agent 2: 108.4875
    For agent 3: 108.4875
    For agent 4: 67.08
    For agent 5: 108.4875
For partition 40 the probability weight is 0.2
The conditional expected value payoffs are
    For agent 1: 108.4875
    For agent 2: 108.4875
    For agent 3: 67.08
    For agent 4: 108.4875
    For agent 5: 108.4875
For partition 44 the probability weight is 0.2
The conditional expected value payoffs are
    For agent 1: 108.4875
    For agent 2: 67.08
    For agent 3: 108.4875
    For agent 4: 108.4875
    For agent 5: 108.4875
For partition 51 the probability weight is 0.2
The conditional expected value payoffs are
    For agent 1: 67.08
    For agent 2: 108.4875
    For agent 3: 108.4875
    For agent 4: 108.4875
    For agent 5: 108.4875

Let us see how this occurs. Since the game is quite symmetrical we need consider only the natural permutation $\{a, b, c, d, e\}$.

(1) At the first stage, agent 1 is added to the null coalition in partition 1, the fine partition $\{\{a\}, \{b\}, \{c\}, \{d\}, \{e\}\}$. The value of each coalition in the fine partition is 51.28.

    (a) Agent b then has the choice of joining with agent a to form a two person coalition in partition 2, $\{\{a, b\}, \{c\}, \{d\}, \{e\}\}$ or remaining independent as a singleton coalition in the fine partition. If he chooses the first option, then by marginality his payoff for this permutation is $v(\{a,b\},2) - v(\{a\},1) = 110.34 - 51.28 = 59.06$. Suppose instead that agent b remains as a singleton at this stage.

(b) Agent c then has three options: remain as a singleton, form coalition $\{a,c\}$, or form coalition $\{b,c\}$. For either of the latter choices the marginal imputation is, again, 59.06.

(c) Suppose instead that agent c remains as a singleton. Then the best possible outcome is that agents d and e join with either a or b and form a three-person coalition. In that case the value of $\{c\}$ is 57.64.

(d) Thus, agent c can gain nothing by free-riding and will choose to form a two-person coalition such as $\{a,c\}$ in partition 3, $\{\{a,c\},\{b\},\{d\},\{e\}\}$. Since they provide the same marginal payment $\{a,c\}$ and $\{b,c\}$ are equally probable, from the point of view of agent b.

  (i) Suppose in particular that partition 6, $\{\{a,c\},\{b\},\{d\},\{e\}\}$, is formed. Then agent d can form a one, two, or three-person coalition. The three-person coalition $\{a,c,d\}$ in partition 11, $\{\{a,c,d\},\{b\},\{e\}\}$, is valued at 209.26, so agent d's marginal payment is $209.26 - 110.34 = 98.92$, and this is his best response.

  (ii) At the last stage agent e can form a four, two, or one-person coalition. The four-person coalition $\{a,c,d,e\}$ in partition 44, $\{\{a,c,d,e\},\{b\}\}$, is valued at 433.95, so agent e's marginal payment is $433.95 - 209.26 = 224.69$, and this is agent e's best response. In this case $\{b\}$ is valued at 67.88, partly thanks to the positive externality from $\{a,c,d,e\}$.

(e) On the other hand, if agent c forms $\{b,c\}$ then agent b's marginal value is the value of the singleton coalition in the fine partition, 51.28.

(2) Thus, if he remains as a singleton, agent b faces two equiprobable outcomes: a payoff of 67.88 as a free rider or a payoff of 51.28 as a member of a four-person coalition. The expected value is 59.58, which is more than the marginal payoff from forming the two-person coalition, 59.06. Therefore, agent b remains as a singleton and either partition 44 or partition 51, $\{\{a\},\{b,c,d,e\}\}$, is formed.

Again, agent b does *not* benefit from being a free-rider. The members of the four-person coalition do better and, if agent b were to join the four-person coalition to form the grand coalition, the marginal valuation at this point would be $951.11 - 433.95 = 517.16$. However, allowing agent b to revise the decision he made at stage 2 would move us from the Shapley world into something more like the world of the core of a cooperative game, and presumably the other four agents must then be allowed to change their decisions and demand more than the marginal payments that have been attributed to them. Marginality, as we find it in the Shapley value, supposes that each agent makes a commitment at a particular point in the permutational order, and the result is averaged over all permutations.

This outcome seems implausible. What does it tell us about the usefulness of marginality in the solution of a game in partition function form? Strictly speaking, the implausible result does not arise from marginality, but more generally, from the assumption that each partition is formed by a sequence of individual decisions, made in a permutational order, and *without any reconsideration*, such as mergers or dissolutions of coalitions previously formed. While it is difficult to define marginality without that presupposition, other forms of valuation based on sequential decisions, such as that proposed by Maskin, would seem to be vulnerable to similar issues.

4.4.7.   *A Game with an Optimal Scale*

We next consider Game 4.6, a four-person game that we shall call the Marshall Game, shown in Table 4.12. This game is not superadditive and is meant to capture, at a very small scale, the idea from Marshallian economics that some kinds of coalitions (business firms) may have an optimal scale. In this case the optimal scale envisioned is 2. Notice that partition 10,

Table 4.12.   Game 4.6: The Marshall game.

1	$\{a\}, \{b\}, \{c\}, \{d\}$	5, 4, 3, 2
2	$\{a, b\}, \{c\}, \{d\}$	12, 2, 1
3	$\{a, c\}, \{b\}, \{d\}$	12, 3, 1
4	$\{a\}, \{b, c\}, \{d\}$	4, 10, 1
5	$\{a, b, c\}, \{d\}$	15, 3
6	$\{a, d\}, \{b\}, \{c\}$	12, 3, 2
7	$\{a\}, \{b, d\}, \{c\}$	4, 10, 2
8	$\{a\}, \{b\}, \{c, d\}$	4, 3, 10
9	$\{a, b, d\}, \{c\}$	15, 3
10	$\{a, b\}, \{c, d\}$	14, 12
11	$\{a, c, d\}, \{b\}$	12, 4
12	$\{a, c\}\{b, d\}$	10, 10
13	$\{a, d\}\{b, c\}$	10, 10
14	$\{a\}\{b, c, d\}$	4, 12
15	$\{a, b, c, d\}$	16

$\{a, b\}\{c, d\}$, generates the largest total value.[1] The computed solution will be given in full. It is given as Table 4.13.

We see that the probability that the efficient partition will form is fairly low, at $\frac{1}{6}$, while there is a substantial probability that inefficient three-person partitions will form and some probability that the inefficient grand coalition will form.

4.4.8.   *A Game with Increasing Returns to Scale*

Game 4.7 is a game of entrepreneurship. As Aumann and Dreze (1974) state, "'acting together' and sharing the proceeds may change the nature of the game." According to John Bates Clark (1899), the formation of a coalition for purposes of production of goods and services changes the nature of the game in that it implies the need for a specialist in the entrepreneurial function of coordination. Game 4.7 is suggested by that idea. In Game 4.7, shown in Table 4.14, Player a is a pure coordinator, who produces nothing in a singleton coalition. The other players are craftsmen who can produce positive amounts working as individual craftsmen in singleton coalitions. However, a coalition of two or more of agents b, c, d produces nothing unless Player a is also a member (since otherwise it would be uncoordinated) but produces an amount that depends only on the number of non-entrepreneur members, supposing Player a is present. In this game there are increasing returns to employment scale, with production (coalition value) of 5 in a coalition containing one craftsman, 15 in a coalition containing 2 craftsmen, or 26 in the grand coalition of three craftsmen together with the entrepreneur. As a result, the grand coalition is the only efficient partition in this game. Nevertheless, the game is not superadditive, and the complexity of its non-superadditive structure is one reason for its interest. Partitions 2–14 are all dominated by *both* partition 1, the fine partition, and by partition 15, the grand coalition. The computed solution is shown in Table 4.15.

We see that in this case, no coalition of two or more agents can be formed, although the grand coalition is Pareto-efficient and unconditionally dominates every other partition. This may not be surprising, given (1) the lack of any intermediate stages not dominated by the fine partition, and (2) the step-by-step procedure of coalition formation assumed by the

---

[1] Strictly speaking, though, it is not uniquely Pareto-optimal. Suppose, for example, that in partition 5, payoffs are 1, 1, 13. Then a shift to partition 10 is not a Pareto-improvement, since agent c can obtain no more than 12 in partition 10.

Table 4.13.   The solution of the Marshall game.

For partition 1 the probability weight is 0
For partition 2 the probability weight is 0
For partition 3 the probability weight is 0
For partition 4 the probability weight is 0
For partition 5 the probability weight is 0.2083333
The conditional expected value payoffs are
    For agent 1: 6
    For agent 2: 5
    For agent 3: 4
    For agent 4: 3
For partition 6 the probability weight is 0
For partition 7 the probability weight is 0
For partition 8 the probability weight is 0
For partition 9 the probability weight is 0.2083333
The conditional expected value payoffs are
    For agent 1: 6.1
    For agent 2: 5.2
    For agent 3: 3
    For agent 4: 3.7
For partition 10 the probability weight is 0.1666667
The conditional expected value payoffs are
    For agent 1: 7.5
    For agent 2: 6.5
    For agent 3: 6.5
    For agent 4: 5.5
For partition 11 the probability weight is 0
For partition 12 the probability weight is 0.1527778
The conditional expected value payoffs are
    For agent 1: 6.363636
    For agent 2: 6.818182
    For agent 3: 3.636364
    For agent 4: 3.181818
For partition 13 the probability weight is 0.125
The conditional expected value payoffs are
    For agent 1: 6.5
    For agent 2: 6.333333
    For agent 3: 3.666667
    For agent 4: 3.5
For partition 14 the probability weight is 0.0694444
The conditional expected value payoffs are
    For agent 1: 4
    For agent 2: 5.7
    For agent 3: 3.5
    For agent 4: 2.8
For partition 15 the probability weight is 0.0694444
The conditional expected value payoffs are
    For agent 1: 4
    For agent 2: 5.7
    For agent 3: 3.5
    For agent 4: 2.8

Table 4.14.   Game 4.7: A John Bates Clark game.

1	$\{a\}, \{b\}, \{c\}, \{d\}$	0, 9, 8, 7
2	$\{a, b\}, \{c\}, \{d\}$	5, 8, 7
3	$\{a, c\}, \{b\}, \{d\}$	5, 9, 7
4	$\{a\}, \{b, c\}, \{d\}$	0, 0, 7
5	$\{a, b, c\}, \{d\}$	15, 7
6	$\{a, d\}, \{b\}, \{c\}$	5, 9, 8
7	$\{a\}, \{b, d\}, \{c\}$	0, 0, 8
8	$\{a\}, \{b\}, \{c, d\}$	0, 9, 0
9	$\{a, b, d\}, \{c\}$	15, 8
10	$\{a, b\}, \{c, d\}$	5, 0
11	$\{a, c, d\}, \{b\}$	15, 8
12	$\{a, c\}\{b, d\}$	5, 0
13	$\{a, d\}\{b, c\}$	5, 0
14	$\{a\}\{b, c, d\}$	0, 0
15	$\{a, b, c, d\}$	26

Table 4.15.   Solution to Game 4.7.

For partition 1 the probability weight is 1
The conditional expected value payoffs are
For agent 1: 0
For agent 2: 9
For agent 3: 8
For agent 4: 7

marginal imputation. The marginal payoff to the first agent to join in a two-person coalition is either 5 or negative, while the payoff to the singleton is no less than 7. No agent can benefit by being first.

4.5.   Chapter Summary

The quasi-Shapley algorithm explored in this chapter is an attempt to extend one particular aspect of the Shapley value to games in partition function form. The particular aspect is that the Shapley value is an average of the marginal values of the agents over all orders in which those marginal contributions may be computed. For the quasi-Shapley algorithm, we allow the agents at each step to decide whether to join an existing coalition or to form a new coalition by remaining a singleton, and these decisions are subgame perfect noncooperative equilibrium decisions. This allows us to compute an imputation of payouts for each partition, and further generates probabilities for the formation of the various partitions. The solutions are plausible in some cases. Positive externalities can generate some inefficiencies, but where the positive externalities are less pronounced, the efficient grand coalition may be the only one with positive probability. Negative externalities also have plausible consequences in some examples. However, the computed solutions are surprising and even implausible in some examples. In a public goods game in which there is no private benefit to a free rider, there is nevertheless a free rider with probability one. In a game with an optimal scale, the probability that the optimal partition will be observed is small, and the probability that excessively large coalitions would be formed is more than half. In a game with increasing returns to scale, in which the grand coalition dominates all others but the fine partition dominates all except the grand coalition, no coalitions are formed and economies of scale are not realized, somewhat as Pigou (1920) predicted. These surprising results arise because coalitions are formed by a series of irreversible individual decisions. This is a necessary condition for marginal valuations to be computed in game theory; but it may not be the appropriate condition for games in partition function form. Further, coalition dominance can play no role. Several implausible results are examples in which a dominant coalition cannot form because of the irreversibility of the individual decisions.

Perhaps the glass is half full. In the actual world, people do sometimes make commitments in response to an expectation about the choices others may make, commitments that leave them worse-off than they would have been had they not been required to make the decision at that particular time. The surprising solutions here reflect that tendency. Further, the marginal valuation scheme proposed here provides payoff schedules and predictions of the formation of partitions that are unique, albeit probabilistic ones. In this, it provides an example of what we may reasonably expect of a value solution for partition function games in which the grand coalition needs not necessarily form. We might hope that future research will produce a solution concept that provides similarly specific predictions, while avoiding the implausible implications of sequential individual decisions and marginal valuation.

Nevertheless, the most important lesson seems to be that a solution concept that can be inconsistent with coalition dominance, as the Shapley value can be even in the case of games in coalition function form, is not a good candidate for a descriptive solution concept to be applied to an economic system in which competition is important. In future chapters we will explore solution concepts in which this difficulty does not arise: the nucleolus and bargaining solutions.

Chapter 5

Extension of the Nucleolus to Nontransferable Utility Games in Partition Function Form

A large proportion of research on cooperative or coalitional games assumes that the game is characterized by transferable utility (TU) and is expressed in coalition function form. In this literature, the solution concepts most often discussed and applied are the core (e.g., Peleg and Sudhölter, 2003) and the Shapley (1953) value. These solution concepts have distinct advantages. The core reflects the conditions of coalitional stability against the threat of withdrawal of some members, but may be null, and if not null, may not uniquely determine the sharing of coalition values among their members. The Shapley value can be computed for any game in appropriate form and is unique, and has some other appealing properties such as marginality. In contrast, Schmeidler's (1969) nucleolus is much less discussed, although, like the Shapley value, it can be computed for any game in appropriate form and is unique, and like the core, it reflects coalitional stability against defection, so that, when the core is not null, the nucleolus is always an element of the core. While it lacks some of the interesting properties of the Shapley value, including marginality, the sensitivity to coalitional deviation may be more important for some applications.

For a game in partition function form, the value of a coalition depends on the partition in which it is embedded, as discussed in the previous chapters. A number of models extend the core and the Shapley value to games in partition function form (e.g., Myerson, 1976; Bolger, 1989; Funaki and Yamato, 1999; Albizuri et al., 2005; Do and Norde, 2007; Koczy, 2007; McQuillin, 2009). Aumann and Dreze (1974) extended the nucleolus to games with coalition structures, and McCain (2009) extended the nucleolus to games in partition function form along similar lines.

For nontransferable utility (NTU) games, each coalition is associated not with a determinate value but with a set of feasible vectors of realized utility for the members of the coalition. It is common to assume that the set of feasible utility vectors has some regularity properties such as compactness and convexity. Several studies extend the core to NTU games, and there are important applications, such as the core of an exchange game (e.g., Debreu and Scarf, 1963; Scarf, 1967). There is also a modest literature on the extension of the Shapley value and related value concepts for NTU games (e.g., Shapley, 1969; Aumann, 1985; see Hart, 2004, for a comparison of three related value concepts). These papers apply to games in coalition function form and tend to rely on an arbitrarily given set of weights that express individual preferences in numerical utility terms, with the computation of the Shapley or other value imputation from the numerical utilities, followed by a test of consistency. Uniqueness may not be assured in general. At least one study (Michener *et al.*, 1989) applies this approach to the nucleolus and Ray and Vohra (1999) remark that some of their results can be extended to some categories of NTU games.

This chapter proposes an extension of the nucleolus to NTU games in partition function form, with unique endogenous weights expressing preferences as numerical utilities. NTU games in partition function form have been discussed in Chapter 3, Section 3.3, and the terminology and notation in this chapter will be as defined there.

5.1. Excess

A key measure in Schmeidler's analysis is the *excess* that a group might generate by deviating from the coalitions they currently participate in. For Schmeidler, the excess is defined for a coalition C and a deviating group S and is the difference $V(S) - \sum_{i \in S} U_i$ where the U_i are evaluated at the current coalitions and imputations. This excess expresses the strength of the deviation's objections to the current coalition structure and payoffs and thus, arguably, it expresses their bargaining power in the given coalition structure. A difficulty will be to evaluate the excess for a deviating group if the summation of utilities is meaningless, as in the Paretian case. Even if the summation is meaningful, i.e., in the intermediate case, restrictions on transfers among members of S after the deviation may mean that not all deviating members are equally committed to the deviation, as the interpretation of the excess suggests they would be.

We suppose that Γ is an NTU game in partition function form such that $f_{C,\mathcal{P}}$ is strictly convex for every embedded coalition $C, \mathcal{P}$. Let $\{U_i^{\mathcal{P}}\}$ be a set of utility indices yielding Pareto optimality for each embedded coalition. For $\mathcal{P} \in \Pi$, a deviation is a set $S \notin \mathcal{P}$. Let σ be the successor function and $\mathcal{R} = \sigma(\mathcal{P}, S)$. Let $\mathcal{P}, \{U_i^{\mathcal{P}}\}$ be a candidate solution. If there is a deviation S and a set $\{U_i^{\mathcal{R}}\}_{i \in S}$ that is Pareto-optimal for $\mathcal{R}, S$; and $\forall m \in S, U_m^{\mathcal{R}} \geq U_m^{\mathcal{P}}$ and moreover $\exists m \in S, U_m^{\mathcal{R}} > U_m^{\mathcal{P}}$, then the candidate solution will be said to be "unstable." Otherwise it is "stable." The core of this game may be defined as the set of stable candidate solutions.

Let $S = \{m, n, \ldots, q\} \subseteq N$ be a deviation from $\mathcal{P}$, and S is the cardinality of the deviation. For the purposes of this chapter we will measure the excess for partition $\mathcal{P}$ and deviation set S as

Definition 5.1.

$$e_{S,\mathcal{P}}(U^{\mathcal{P}}) = \min_{i \in S}(U_i^{\mathcal{R}} - U_i^{\mathcal{P}})$$

where $U_i^{\mathcal{R}}$ is the utility that agent i can expect in partition $\mathcal{R}$. The set of imputations $\{U_i^{\mathcal{P}}\}_{i \in C}$ will be denoted as $U^{\mathcal{P}}$, $\{U_i^{\mathcal{R}}\}_{i \in C}$ as $U^{\mathcal{R}}$, and so on. For the present, $U^{\mathcal{R}}$ will be taken as given. Thus, the bargaining power of the deviation is identified with the "weakest link" of the deviation. The deviation is a cooperative move by the members of S, and thus its plausibility as a threat is identified with the benefit to the agent who gains the least, and so is most likely to veto the move. Note that, if $e_{S,\mathcal{P}}$ is positive, every member of S has $U_i^{\mathcal{R}} > U_i^{\mathcal{P}}$ and so the coalition is unstable.

5.2.   The Nucleolus

Let $\Xi_C = \{S \subseteq N \ni S \neq C, \exists i \in C \cap S\}$; that is, Ξ_C is the set of all deviations from $\mathcal{P}$ in which members of C might participate. Let $j \in C$. Then $\Xi_j = \{S \subseteq n \ni j \in S, S \notin \mathcal{P}\}$; that is, the set of all deviations from $C, \mathcal{P}$ in which agent j might participate. Note that $\Xi_j \subseteq \Xi_C$.

Let Γ be an NTU game in partition function form.

Now consider the following *noncooperative* game in strategic normal form, Γ^*. The *players* in Γ^* are the embedded coalitions in Γ. Notice that a particular coalition will enter as a separate player in each partition of which it is an element. The strategies for each embedded coalition $\mathcal{P}, C$ are the imputations $U_i^{\mathcal{P}}$, $i \in C$. The payoff for $C \in \mathcal{P}$ is the negation,

$$- \max_{S \in \Xi_C}(e_{S,\mathcal{P}}(U^{\mathcal{P}})) \tag{5.1a}$$

that is,

$$- \max_{S \in \Xi_C} \min_{i \in S}(U_i^{\mathcal{R}} - U_i^{\mathcal{P}}). \tag{5.1b}$$

Accordingly, the best response for C will be to

$$\min_{U_i^{\mathcal{P}}} \max_{S \in \Xi_C} \min_{i \in S}(U_i^{\mathcal{R}} - U_i^{\mathcal{P}}). \tag{5.1c}$$

Note that S is a deviation from the partition $\mathcal{P}$. Now, $S \in \Xi_C$, so it follows that $C \cap S \neq \varnothing$. For a coalition $B \in \mathcal{P}$, $B \neq C$, it may be that $B \cap S = \varnothing$. In such a case there is no redistribution of utility assignments among the members of B that could influence the excess of S. Accordingly, the utility assignment of B will reflect only those deviations T from $\mathcal{P}$ whose excesses B can influence, i.e., those for which $B \cap T \neq \varnothing$. However, $U_i^{\mathcal{R}}$ is the strategy of whichever coalition in $\mathcal{R}$ contains i, which may not be S.

Since the strategies are drawn from a compact set, there is at least one Nash equilibrium in pure strategies. We now consider some properties of the equilibrium, for the case of a strictly convex UPF.[1]

Let i, j be such that

$$i \in C,\ j \in C,\ \mathcal{R} - \sigma(\mathcal{P}, C),\ i \subset S_1 \subseteq \mathcal{R},\ j \in S_2 \in \mathcal{R},$$
$$(U_i^{\mathcal{R}} - U_i^{\mathcal{P}}) = \min_{U_i} \max_{S \in \Xi_C}(e_{\mathcal{P},S}(U_i^{\mathcal{P}})) = \min_{U_i} \max_{S \in \Xi_S} \min_{i \in S}(U_i^{\mathcal{R}} - U_i^{\mathcal{P}}); \tag{5.2}$$

but suppose that nevertheless $(U_j^{\mathcal{R}} - U_j^{\mathcal{P}}) > (U_i^{\mathcal{R}} - U_i^{\mathcal{P}})$.

Consider the total differential of $e_{\mathcal{P},S}(U^{\mathcal{P}})$ with respect to the utilities of i and j,

$$de = \frac{\partial e_{\mathcal{P},S}(U_i^{\mathcal{P}})}{\partial U_i^{\mathcal{P}}} dU_i + \frac{\partial e_{\mathcal{P},S}(U_i^{\mathcal{P}})}{\partial U_j^{\mathcal{P}}} dU_j. \tag{5.3}$$

Supposing that $dU_i^{\mathcal{P}} > 0$ and $dU_j^{\mathcal{P}} < 0$; that is, supposing we transfer utility from j to i along the utility possibility frontier. Since $\frac{\partial e_{\mathcal{P},S}(U_i^{\mathcal{P}})}{\partial U_j^{\mathcal{P}}} dU_j = 0$, that is

$$de = \frac{\partial e_{\mathcal{P},S}(U_i^{\mathcal{P}})}{\partial U_i^{\mathcal{P}}} dU_i < 0. \tag{5.4}$$

---

[1]Note, there is no assumption here that the values of coalitions are influenced by noncooperative play among the coalitions. Rather, the formalism of the Nash equilibrium is used simply to characterize a consistent set of conditional maxima.

That is, the original imputation $U_i^\mathcal{P}$, $U_j^\mathcal{P}$ was not a best response to the imputation strategies of S. The excess for S can be reduced by a transfer from j to i. Thus, U_j as in (5.2) would not constitute a best response to $U_j^\mathcal{R}$ and instead, at a best response, for $j \in S \cap C$, $\ni U_j^\mathcal{R} - U_j^\mathcal{P} \leq \min_{U_i} \max_{S \in \Xi_C}(e_{S,\mathcal{P}}(U_i^\mathcal{R}))$. Since by the definition of the excess, $U_j^\mathcal{R} - U_j^\mathcal{P} \geq \min_{U_i} \max_{S \in \Xi_C}(e_{S,\mathcal{P}}(U_i^\mathcal{P}))$, we have $U_j^\mathcal{R} - U_j^\mathcal{P} = \min_{U_i} \max_{S \in \Xi_C}(e_{S,\mathcal{P}}(U_i^\mathcal{P}))$.

In more formal terms, the best response for coalition $C \in \mathcal{P}$ will satisfy a series of maximization and minimization programs. Take $U_j^\mathcal{R}$ as given for all $j \in C$, determined by the strategies of the coalitions in $\mathcal{R}$. From (5.1a), we have

$$\max_{S \in \Xi_C} e_{\mathcal{P},S}(U^\mathcal{P}) \tag{5.5a}$$

that is, equivalently

$$\max_{S \in \Xi_C} \min_{i \in S}(U_i^\mathcal{R} - U_i^\mathcal{P}). \tag{5.5b}$$

The solution of this program needs not be unique and in general will not be. Since, however, the values are drawn from a finite set, there must be at least one solution. Let S, i be a solution to (5.5b) and $\mathcal{R} = \sigma(\mathcal{P}, S)$. Then we have, for $j \neq i$,

$$U_j^\mathcal{R} - U_j^\mathcal{P} \geq U_i^\mathcal{R} - U_i^\mathcal{P} \tag{5.6a}$$

$$U^\mathcal{R} - U^\mathcal{P} \geq e(\mathcal{P}, T, U_i^\mathcal{P}) \quad \forall T \in \Xi_C, \ T \neq S. \tag{5.6b}$$

The first inequality applies for all $j \in C$, $j \neq i$, and the second applies for all $T \in \mathcal{P}$, $T \neq C$. The best response must also solve

$$\min_{U_i^\mathcal{P}} U_i^\mathcal{R} - U_i^\mathcal{P} \tag{5.5c}$$

subject to (5.6a), (5.6b) and also to

$$f_{C,\mathcal{P}}(\{U_j^\mathcal{P}\}_{j \in C}) \leq 0, \tag{5.6c}$$

the feasibility constraint.

Thus we form the Lagrangean function

$$\mathcal{L} = U_i^\mathcal{R} - U_i^\mathcal{P} + \mu f_{C,\mathcal{P}}(\{U_j^\mathcal{P}\}_{j \in C}) + \sum_{\substack{j \in C \\ j \neq i}} \lambda_j [U_j^\mathcal{P} - U_j^\mathcal{Q} - (U_i^\mathcal{P} - U_i^\mathcal{R})]$$

$$+ \sum_{\substack{T \in \mathcal{P} \\ T \neq C}} \lambda_T [U_i^\mathcal{R} - e_{\mathcal{P},T}(U^\mathcal{P})]. \tag{5.7a}$$

Setting

$$\lambda_i = 1 - \sum_{\substack{j \in C \\ j \neq i}} \lambda_j + \sum_{T \in \Xi_C} \lambda_T \qquad (5.7b)$$

and noting that $U_i^{\mathcal{R}}$ and $e_{\mathcal{P},T}(U^{\mathcal{P}})$ are constants for the computation of the best response in the constructed noncooperative game, this is equivalent to the *maximization* program corresponding to the Lagrangean function

$$\mathcal{L} = \sum_{j \in C} \lambda_j U_j^{\mathcal{P}} - \mu f_{C,\mathcal{P}}(\{U_j\}_{j \in C}). \qquad (5.7c)$$

Provided $\lambda_j > 0$, this characterizes a Pareto optimum. The first-order necessary conditions for a solution are

$$\frac{\partial \mathcal{L}}{\partial U_j^{\mathcal{P}}} = \lambda_j - \mu \frac{\partial f_{C,\mathcal{P}}}{\partial U_j} \leq 0 \qquad (5.7d)$$

so that, for an interior solution, the Pareto optimality, within each coalition, of the Nash equilibrium is assured.

If we impose a non-negativity constraint on U_j, then we might have a corner solution. If, however, we interpret the utility indices as von Neumann–Morgenstern utilities, nothing can depend on the zero of the utility scale, so that a non-negativity constraint would be inappropriate. If the utility possibility set is strictly convex, then the maximum is unique.

Lemma 5.1. *Denoting $U_j^{\mathcal{P}}$ and $U_j^{\mathcal{R}}$ as before, $\forall j \in C$, $U_j^{\mathcal{R}} - U_j^{\mathcal{P}} = e_{\mathcal{P},S}(U^{\mathcal{P}})$.*

Proof. By the definition of the excess, $U_j^{\mathcal{R}} - U_j^{\mathcal{P}} \geq e_{\mathcal{P},S}(U^{\mathcal{P}}) = U_i^{\mathcal{R}} - U_i^{\mathcal{P}}$. Suppose $U_j^{\mathcal{R}} - U_j^{\mathcal{P}} > U_i^{\mathcal{R}} - U_i^{\mathcal{P}}$. Then a redistribution from j to i will reduce $U_i^{\mathcal{R}} - U_i^{\mathcal{P}}$, contrarily to (5.5c). Thus $U_j^{\mathcal{R}} - U_j^{\mathcal{P}} = U_i^{\mathcal{R}} - U_i^{\mathcal{P}}$. Since furthermore $U_i^{\mathcal{R}} - U_i^{\mathcal{P}} = e_{\mathcal{P},S}(U^{\mathcal{P}})$,

$$U_j^{\mathcal{R}} - U_j^{\mathcal{P}} = e_{\mathcal{P},S}(U^{\mathcal{P}}) \quad \forall j \in C. \qquad (5.8)$$

$\square$

Lemma 5.2. *For an NTU game in partition function form Γ, $\mathcal{P} \in \Pi_N$, $C \in \mathcal{P}, \exists S \subseteq N, S \notin \mathcal{P}$ and $\exists y_C$ a constant $\ni \forall i \in C, U_i^{\mathcal{P}} = U_i^{\mathcal{R}} + y_C$ where $\mathcal{R} = \sigma(\mathcal{P}, S)$.*

Proof. S will correspond to (5.5a), so that $U_i^{\mathcal{R}} - U_i^{\mathcal{P}} = e_{\mathcal{P},S}(U^{\mathcal{P}})$ and $e_{\mathcal{P},S}(U^{\mathcal{P}}) = y_C$. $\square$

The nucleolus for game Γ, then, comprises for each partition $\mathcal{P}$ the utility assignments corresponding to the Nash equilibrium of the game Γ^*. By (5.8), each member of a coalition receives what he would obtain in the most profitable deviation from the coalition, plus an equal amount from the surplus generated by the coalition (or minus an equal assessment for its losses). Note that if $\mathcal{P}$ is stable, then for each $\mathcal{C} \in \mathcal{P}$, $\min_{U_i} \max_{T \in \Xi_C} (e_{T,\mathcal{P}}(U^{\mathcal{P}}))$ is negative and so the imputation satisfies all rationality constraints including individual rationality, while if $\mathcal{P}$ is unstable, it does not.

The Nash equilibrium formalism provides a direct extension of Schmeidler's solution concept. Schmeidler defines a lexicographic ordering, ordering deviation sets in terms of increasing excesses and payoff vectors in order, first, of the deviation sets, and second, of the excess for the corresponding deviation set, and identifies the nucleolus with the minimum according to this ordering. The idea is that the compensation committee wishes to distribute payouts so as to reduce discontent to the greatest extent possible. Thus, they would prefer to reduce the discontent of the most discontented potential deviation. Then, if they are unable to do so, they consider the second most discontented, and so on, until they distribute the payouts so as to minimize the discontent of the most discontented group that they can effectively benefit.

5.3. Uniqueness

Lemma 5.3. *The nucleolus is unique.*

Proof.   Suppose there are two distinct Nash equilibria of Γ^*. Then $\exists \mathcal{P} \in \Pi_N$, $i \in S \ni {}^1U_i^{\mathcal{P}}$, ${}^2U_i^{\mathcal{P}}$ correspond to Nash equilibria of Γ^*. Suppose ${}^1U_i^{\mathcal{P}}$ is computed for a deviation to S and suppose ${}^2U_i^{\mathcal{P}}$ is computed for a deviation to T. By (5.5a) we have both

$$e_{P,S}({}^1U^{\mathcal{P}}) \geq e_{P,T}({}^1U^{\mathcal{P}})$$
$$e_{P,T}({}^2U^{\mathcal{P}}) \geq e_{P,S}({}^2U^{\mathcal{P}}). \tag{5.9a}$$

Using (5.8),

$$U_j^{\mathcal{R}} - {}^1U_j^{\mathcal{P}} \geq U_j^{\mathcal{Q}} - {}^1U_j^{\mathcal{P}} \quad \text{for } \mathcal{R} = \sigma(\mathcal{P}, S), \ \mathcal{Q} = \sigma(\mathcal{P}, T), \tag{5.9b}$$

therefore

$$U_j^{\mathcal{Q}} \geq U_j^{\mathcal{R}}. \tag{5.9c}$$

By similar reasoning,

$$U_j^{\mathcal{R}} \geq U_j^{\mathcal{Q}}. \tag{5.9d}$$

Therefore,

$$U_j^{\mathcal{R}} = U_j^{\mathcal{Q}}. \tag{5.9e}$$

Without loss of generality, suppose ${}^1U_i^{\mathcal{P}} > {}^2U_i^{\mathcal{P}}$. Then

$$U_i^{\mathcal{Q}} - {}^2U_i^{\mathcal{P}} = e_{P,T}({}^2U^{\mathcal{P}}) \geq U_i^{\mathcal{R}} - {}^2U_i^{\mathcal{P}} > U_i^{\mathcal{R}} - {}^1U_i^{\mathcal{P}} \tag{5.9f}$$

which contradicts the minimization at (5.5c) for the computation of ${}^2U_i^{\mathcal{P}}$. Thus, while there might be plural equilibria in the sense that more than one deviation might satisfy (5.5a), all will give rise to the same values for the nucleolus. □

We will now formalize some of the results from this section and the previous one.

Theorem 5.1. *For a NTU game in partition function form, such that $f_{C,\mathcal{P}}$ defines a strictly convex utility possibility set for each embedded coalition, for each $\mathcal{P} \in \Pi$, an imputation consistent with (5.1c) exists and is unique and, for each $C \in \mathcal{P}$, $\exists S \notin \mathcal{P}$, for each $i \in C$, the imputation takes the form $U_i^{\mathcal{R}} + y_C$ where y_C is a constant for all $i \in C$ and $U_i^{\mathcal{R}}$ is the value of i in $\mathcal{R} = \sigma(\mathcal{P}, S)$.*

The Nash equilibrium will thus specify a unique distribution of utilities U_i^* for each partition $\mathcal{P}$ and a corresponding set of distributional weights λ_i^*; and thus, implicitly, a value for the coalition, $v(\mathcal{P}, C) = \sum_{i \in C} \lambda_i^* U_i$. For partition $\mathcal{R} = \sigma(\mathcal{P}, S)$, let β_i be the Nash-equilibrial distributional weights. Since for $i \in S$, we have $U_i^{\mathcal{R}} - U_i^{\mathcal{P}} = e_{P,S}(U^{\mathcal{P}})$ it follows that

$$v(\mathcal{R}, S) - \sum_{i \in S} \beta_i U_i = e_{P,S}(U^{\mathcal{R}}) \sum_{i \in S} \beta_i. \tag{5.10}$$

Normalizing so that $\sum_{i \in S} \beta_i = 1$, we have

$$e_{P,S}(U^{\mathcal{P}}) = v(\mathcal{R}, S) - \sum_{i \in S} \beta_i U_i^{\mathcal{P}}. \tag{5.11}$$

That is, the excess attributable to deviation S is the value of the deviation in the successor partition net of the utilities attributed to its members in the original partition, measured at the weights for the successor partition

5.4. Characterizing the Core

While this chapter is primarily concerned with the nucleolus, it will be helpful to discuss the core in a more careful way than some passing references so far, in order to clarify the relationship of the nucleolus to the core. As before, let $U_i^{\mathcal{P}}$ be an assignment of utilities admissible for partition $\mathcal{P}$, $S \notin \mathcal{P}$ a deviation, $\mathcal{R} = \sigma(\mathcal{P}, S)$, and $f_{\mathcal{R},S}$ the constraint function for S conditional on partition $\mathcal{R}$. If

$$f_{\mathcal{R},S}(U^{\mathcal{P}}) < 0, \tag{5.12a}$$

then S will be able to offer

$$\{U_i^{\mathcal{R}}\}_{i \in S} = \{U_i^{\mathcal{P}} + \varepsilon_i\}_{i \in S} \tag{5.12b}$$

with $\varepsilon_i > 0$ and

$$f_{\mathcal{R},S}(U^{\mathcal{R}}) \leq 0, \tag{5.12c}$$

and so $\mathcal{P}, U^{\mathcal{P}}$ will be unstable in the sense that an imputation not in the core is unstable. Conversely, $\mathcal{P}, U^{\mathcal{P}}$ will be stable in that sense if, for any $S \notin \mathcal{P}$, $f_{\mathcal{R},S}(\{U_i^{\mathcal{P}}\}_{i \in S}) \geq 0$. (They will be strictly stable if $f_{\mathcal{R},S}(\{U_i^{\mathcal{P}}\}_{i \in S}) > 0$.) Thus, given Γ an NTU game in partition function form and $\mathcal{P} \in \Pi_N$, the core for $\mathcal{P}$ comprises all feasible utility assignments $U_i^{\mathcal{P}}$ satisfying the condition that, for $S \subseteq N$, $S \notin \mathcal{P}$, $f_{\mathcal{R},S}(\{U_i^{\mathcal{P}}\}_{i \in S}) \geq 0$.

Theorem 5.2. *If the core is non-null for $\mathcal{P}$, then the nucleolus is an element of the core.*

Proof.   Let U^n denote the nucleolus for $\mathcal{P}$ and let $U^{\mathcal{P}}$ be an element of the core for $\mathcal{P}$. By the definition of excess, $e_{\mathcal{P},S}(U^{\mathcal{P}}) \leq U_i^{\mathcal{R}} - U_i^{\mathcal{P}}$, $\forall i \in S$, moreover, $e_{\mathcal{P},S}(U^n) \leq e_{\mathcal{P},S}(U^{\mathcal{P}})$. Thus, using (5.8),

$$U_i^{\mathcal{R}} - U_i^n \leq U_i^{\mathcal{R}} - U_i^{\mathcal{P}} \tag{5.13a}$$

that is,

$$U_i^{\mathcal{P}} \leq U_i^N. \tag{5.13b}$$

Thus, from the properties of the constraint function,

$$f_{\mathcal{R},S}(\{U^n\}_{i \in S}) \geq f_{\mathcal{R},S}(\{U^{\mathcal{P}}\}_{i \in S}) \geq 0. \tag{5.13c}$$

Thus far we have characterized the core for a particular coalition structure, along lines suggested by Aumann and Debreu (1974). We may extend the concept as follows. A candidate solution for an NTU game Γ in partition function form will be a partition $\mathcal{P}$ and a utility assignment U_i admissible for $\mathcal{P}$. The candidate solution will be stable if U_i is an element of the core for $\mathcal{P}$. Then we identify the core for Γ as the set of all pairs $(\mathcal{P}, U_i)$ such that U_i is an element of the core for $\mathcal{P}$. Moreover the set $\{\mathcal{P}, U^{\mathcal{P}} | \mathcal{P} \text{ is stable}\}$ is the core for $\mathcal{P}$.   □

5.5.   The Intermediate Case and Proper Games

Consider first the intermediate case with $f_{C,\mathcal{P}}$ strictly convex for all embedded coalitions. The interpersonally comparable cardinal utilities might be computed along the lines of Fellner (1967). Supposing that consumer preferences are additively separable in two groups of commodities, e.g., $x_1 = $ foodstuffs and $x_2 = $ non-food, then $U_j = U_j(x_1) + U_j(x_2)$, which can be identified from demand data for x_1 and x_2. Moreover, as food is a basic need, we might regard the utility of the two individuals as comparable in that their units reflect the degree to which the two have met the common, human basic need for food. Nevertheless, if $f_{C,\mathcal{P}}$ is strictly convex, $\sum_{i \in C} U_i$ will not be a constant and thus cannot be identified with $V_{C,\mathcal{P}}$. Thus, as before, we identify the nucleolus with the Nash equilibrium of game Γ^*, the equilibrium $U_i^{\mathcal{P}}$ and λ_i, and $V_{C,\mathcal{P}} = \sum_{i \in C} \lambda_i U_i^{\mathcal{P}}$.

But suppose instead that Γ is a proper game. Then Eq. (5.1c) is satisfied for any weights λ_i, provided $\sum_{i \in C} U_i^{\mathcal{P}} = V_{C,\mathcal{P}}$. In particular it is satisfied for weights $\lambda_i = 1$. Thus Eq. (5.13) reduces to Schmeidler's definition of the excess, and his procedure minimizes the maximum excess as in Eq. (5.1a). In this sense, Schmeidler's nucleolous is an instance of the nucleolus as defined here, for the special case of a proper game.

5.6.   Extended Core

In the definition of the core in Sections 5.1 and 5.4, this discussion followed the convention by defining instability for a candidate solution along these lines: If there is a deviation S from $\mathcal{P}$ and a feasible set $\{U_i^S\}$ such that some $i \in S$ are better off, and none worse off, then the partition $\mathcal{P}$ is unstable. It may be, however, that the payoffs $\{U_i^S\}$ do not correspond to the nucleolus for the successor partition, and that some members of S are in fact worse off if the utilities in the successor partition are assigned according to the nucleolus. Then those agents would veto the deviation: this is to say, the partition may be stable despite the fact that all the members of a deviation can in principle benefit by the deviation; but nevertheless some members rationally expect that their bargaining power would be so compromised by the deviation that the deviation does not take place. We might then define an extended core as follows: Let σ be the successor function and $\mathcal{R} = \sigma(\mathcal{P}, \mathcal{S})$. If there is a deviation $\mathcal{S}$ and a set $\{U_i^{\mathcal{R}}\}_{i \in \mathcal{S}}$ such that the $U_i^{\mathcal{R}}$ correspond to the nucleolus for $\mathcal{R}$; and $\forall m \in S$, $U_m^{\mathcal{R}} \geq U_m^{\mathcal{P}}$ and moreover $\exists m \in S \ni U_m^{\mathcal{R}} > U_m^{\mathcal{P}}$, then the candidate solution will be said to be "unstable." Otherwise it is "semi-stable." The extended core of this game may be defined as the set of semi-stable candidate solutions. Clearly the extended core is a superset of the core.

5.7. Superior Games

Superior games have been introduced in Chapter 3, Section 3.3.1, and what follows is consistent with the discussion here. We now consider some implications of superiority for the nucleolus.

Lemma 5.4. *For a superior game* Γ, $\mathcal{P} \in \Pi_N$, $C \in \mathcal{P} \neq \mathcal{G}$, $e_{\mathcal{P},N}(U_i^{\mathcal{P}}) = \max_{S \in \Xi_C} e_{\mathcal{P},S}(U_i^{\mathcal{P}})$.

Proof. Suppose $T \in \Xi_C \ni e_{\mathcal{P},T}(U_i^{\mathcal{P}}) > e_{\mathcal{P},N}(U_i^{\mathcal{P}})$. Let $\mathcal{R} = \sigma(\mathcal{P}, T)$. Then $\forall j \in T$, $U_i^{\mathcal{R}} - U_i^{\mathcal{P}} > e_{\mathcal{P},N}(U_i^{\mathcal{P}}) \geq U_i^{\mathcal{G}} - U_i^{\mathcal{P}}$; thus $U_i^{\mathcal{R}} > U_i^{\mathcal{G}}$. Moreover, for $B \in \mathcal{R}$, $f_{B,\mathcal{R}}(\{U_j^{\mathcal{R}}\}_{j \in B}) \leq 0$, so that, by Definition 3.17, $f_{\mathcal{G},N}(\{U_j^{\mathcal{R}}\}_{j \in B}) \leq 0$. However, it follows that $f_{\mathcal{G},N}(\{U_j^{\mathcal{R}}\}_{j \in B}) < 0$, which would contradict the supposition that $U_i^{\mathcal{G}}$ is a best response for N as an embedded coalition. $\square$

Remark. Here we treat the grand coalition as a deviation from $\mathcal{P}$ and find that, for a superior game, it is the most profitable deviation.

Lemma 5.5. *For a superior game* Γ, *if* $\mathcal{P} \in \Pi_N$, $\mathcal{P} \neq \mathcal{G}$ *and* $i \in C \in \mathcal{P}$, *letting* $U_i^{\mathcal{P}}$ *denote the nucleolus for* $\mathcal{P}$, $U_i^{\mathcal{P}} = U_i^{\mathcal{G}} + y_C$, *where* y_C *is a constant for all* $i \in C$.

Proof. Consider $\mathcal{G}$ as a deviation from $\mathcal{P}$. Then $e_{\mathcal{P},N}(U_i^{\mathcal{P}}) = \max_{S \in \Xi_C} e_{\mathcal{P},S}(U_i^{\mathcal{P}})$ from Lemma 5.4 and so, by Lemma 5.2, $U_i^{\mathcal{P}} = U_i^{\mathcal{G}} + y_C$, where y_C is a constant for all $i \in C$. $\square$

Theorem 5.3. *If Γ is a strictly superior NTU game in partition function form, then no candidate solution based on any partition other than the grand coalition $\mathcal{G} = \{N\}$ is stable. Moreover, no candidate solution based on a partition other than the grand coalition is quasi-stable.*

Proof. Let $\mathcal{P} \neq \mathcal{G}$ be a partition, $U_i^{\mathcal{P}}$ a feasible imputation for $\mathcal{P}$. For $\mathcal{C} \in \mathcal{P}$, consider $\mathcal{G}$ as a deviation from $\mathcal{C}, \mathcal{P}$. Note that, since the deviating coalition in this case is N and $\mathcal{G} = \{N\}$ is the only partition with N as an element, necessarily $\sigma(\mathcal{P}, N) = \mathcal{G}$. (McCain, 2009, p. 184). From Definition 3.15, $f_{N,\mathcal{G}}(\{U_i^{\mathcal{P}}\}) < 0$. Thus, $\exists\, U_i^{\mathcal{G}} \ni$

(a) $\forall i \in N,\ U_i^{\mathcal{G}} \geq U_i^{\mathcal{P}}$.
(b) $\exists j \in N \ni U_i^{\mathcal{G}} > U_j^{\mathcal{P}}$.
(c) $f_{N,\mathcal{G}}(\{U_i^{\mathcal{G}}\}) \leq 0$.

Letting $\mathcal{C} \in \mathcal{P}$ be such that $j \in \mathcal{C}$, this establishes the instability of $\mathcal{P}, U_i^{\mathcal{P}}$.

$\square$

Now, in particular let $\{U_i^{\mathcal{P}}\}$ be the nucleolus for $\mathcal{P}$ and let $\{U_i^{\mathcal{G}}\}$ denote the nucleolus for $\mathcal{G}$. To show that $\mathcal{P}$ is not quasi-stable it is enough to show that for at least one $i \in N$, $U_i^{\mathcal{G}} > U_i^{\mathcal{P}}$. Suppose not, i.e., suppose that $\forall i \in N,\ U_i^{\mathcal{G}} \leq U_i^{\mathcal{P}}$. Moreover, for $\mathcal{C} \in \mathcal{P}$, $f_{\mathcal{P},\mathcal{C}}(\{U_i^{\mathcal{P}}\}_{i \in \mathcal{C}}) = 0$. Again using Definition 3.15, it would follow that $f_{\mathcal{G},N}(\{U_i^{\mathcal{G}}\}_{i \in N}) < 0$, contradicting the supposition that $U_i^{\mathcal{G}}$ is a best response for N as an embedded coalition.

Remarks. In case $\mathcal{G}$ is itself not stable, then we have dominance cycles and an "empty-core" game in the NTU, partition function framework.

5.8. The Example of a Common-Property Resource

We now revisit an example introduced in Chapter 3, Section 3.2, in which N agents make use of a common-property resource subject to a linear-quadratic benefit function. To determine the excess and thus the nucleolus we require a successor function. Since the game is superior, any rational successor function will be characterized as follows: Suppose that $i \in C_k$ deviates from $\mathcal{P}$ to form a singleton coalition. Then rational reorganization will lead to the consolidation of the remaining members of C_k and the other coalitions to form partition $\mathcal{Q} = \{N \backslash i, \{i\}\}$. However, for some purposes, the naïve successor function may be of interest. The naïve successor function would make partition $\mathcal{R} = \{C_1, \ldots, C_k \backslash i, \{i\}, C_{k+1}, \ldots, C_m\}$ the successor of $\mathcal{P}$ when $\{i\}$ deviates. Following the proposal of Thrall and Lucas (1963), the successor would be the fine partition $F = \{\{1\}, \ldots, \{n\}\}$, since this would be least favorable for the deviating singleton. This proposal seems to have been suggested by the assurance principle applied in most discussions of games in coalition function form, but strictly speaking, the assurance principle would require that each agent use an infinite quantity of the common-property resource, thus reducing the payoffs of all other agents to minus infinity, so that minus infinity would be the value of every coalition other than the grand coalition.

Assuming that the successor function is rational, we will have

$$Y_i^* = y_i + \sum_{\substack{j=1 \\ j\neq i}}^{N} y_j = \sum_{i=1}^{N} \frac{\alpha_i - \gamma_i}{\beta_i} - \sum_{\substack{j=1 \\ j\neq i}}^{N} \frac{1}{\beta_j} \sum_{\substack{s=1 \\ s\neq i \\ s\neq j}}^{N} \frac{\lambda_s}{\lambda_j} \gamma_s \qquad (5.14a)$$

and the deviator's utility will be

$$u_i^* = \alpha_i \left(\frac{\alpha_i - \gamma_i}{\beta_i} \right) - \frac{\beta_i}{2} \left(\frac{\alpha_i - \gamma_i}{\beta_i} \right)^2 - \gamma_i \sum_{i=1}^{N} \frac{\alpha_i - \gamma_i}{\beta_i}$$

$$+ \gamma_i \sum_{\substack{j=1 \\ j\neq i}}^{N} \frac{1}{\beta_j} \sum_{\substack{s=1 \\ s\neq i \\ s\neq j}}^{N} \frac{\lambda_s}{\lambda_j} \gamma_s = u_i^{\mathcal{F}} + \gamma_i \sum_{\substack{j=1 \\ j\neq i}}^{N} \frac{1}{\beta_j} \sum_{\substack{s=1 \\ s\neq i \\ s\neq j}}^{N} \frac{\lambda_s}{\lambda_j} \gamma_s. \qquad (5.14b)$$

The deviator's payout will be increased *both* by the deviator's choice of the dominant strategy (which corresponds to the first three terms) *and* by the rational consolidation of the other players into a single coalition (the last term), which reduces the rate of exploitation of the members of the newly formed coalition $N \setminus \{i\}$. Now, for the nucleolus for partition $\mathcal{P}$, the individual utility will be

$$u_i = u_i^* + B_k \qquad (5.14c)$$

where B_k is a dividend constant over the members of C_k. In this case, the dividend will be

$$B_k = -\frac{\gamma_i h_i}{\beta_i} - \frac{h_i^2}{2\beta_i} - \gamma_i \sum_{\substack{j=1 \\ j\neq i}}^{N} \frac{1}{\beta_j} \sum_{\substack{s=1 \\ s\neq i \\ s\notin C_i}}^{N} \frac{\lambda_s}{\lambda_j} \gamma_s. \qquad (5.14d)$$

For given parameters, this will be a complicated polynomial in the distributional weights, and solution will be beyond the scope of this discussion. We do note, however, that all three terms are negative, which reflects the fact that every other coalition is dominated by some imputations feasible in the grand coalition.

Remark. From this point on we will consider the symmetrical case, i.e., for all $i, j, \alpha_i = \alpha_j = \alpha, \beta_i = \beta_j = \beta, \gamma_i = \gamma_j = \gamma$.

For the symmetrical case, we will have

$$y_i = \frac{1}{\beta} \left(\alpha - \frac{\gamma m_k}{\lambda_i N} \right) \tag{5.15a}$$

$$Y = N \frac{\alpha}{\beta} - \frac{\gamma}{\beta} \sum_{C_k \in \mathcal{P}} \frac{m_k}{N} \frac{1}{H_k} \tag{5.15b}$$

where

$$H_k = \frac{1}{\sum_{j \in C_k} \frac{1}{\lambda_j}} \tag{5.15c}$$

Remark. Thus, H_k is the inverse of the harmonic mean of the distributional weights for coalition C_k. Moreover, the arithmetic mean of the distributional weights for any coalition, given the normalization at (3.16a), is $1/N$; so (5.15b) can be rewritten as

$$Y = N \frac{\alpha}{\beta} - \frac{\gamma}{\beta} \sum_{C_k \in \mathcal{P}} m_k \frac{A_k}{H_k} \tag{5.15d}$$

with A_k denoting the arithmetic mean. Recalling that the harmonic mean is no more than the arithmetic mean, and approaches the arithmetic mean from below as the values averaged approach equality, we may think of A_k/H_k as an index of the inequality of the distributional weights for the coalition. Thus, we conclude that the overall surplus of the coalition decreases as the size of coalitions increases and as the inequality of the distributional weights increases. Intuitively, however, we might expect that the weights will be equal in the symmetrical case, for the nucleolus imputation, and that will prove to be correct.

For individual i, the payoff will be $u_i^* + B_k$ for $i \in C_k$, as noted in (5.14c). Then

$$B_k = -\frac{\gamma}{\beta} h_i - \frac{1}{2\beta} h_i^2 - \frac{\gamma^2}{\beta} \sum_{\substack{j=1 \\ j \notin i}}^{N} \sum_{\substack{s=1 \\ s \neq j \\ s \in C_i}} \frac{\lambda_s}{\lambda_j}. \tag{5.16a}$$

Without solving this quadratic in h_i, we may observe that the solution will be the same for all members of coalition C_k. Denote that solution as h_k. Then, for $i \in C_k$,

$$\lambda_i = \frac{\gamma}{h_k}\left(\frac{m_k}{N} - 1\right) \tag{5.16b}$$

$$y_i = \frac{1}{\beta}(\alpha - \gamma m_k) \tag{5.16c}$$

$$Y^{\mathcal{P}} = N\frac{\alpha}{\beta} - \frac{\gamma}{\beta}\sum_{C_k \in \mathcal{P}} m_k^2. \tag{5.16d}$$

For the grand coalition

$$Y^{\mathcal{G}} = N\left(\frac{\alpha}{\beta} - \frac{\gamma}{\beta}N\right). \tag{5.16e}$$

Then the individual payoff for partition $\mathcal{P}$ will be

$$u^{\mathcal{P}} = \frac{\alpha - \gamma m_k}{\beta} - \frac{1}{2\beta}(\alpha - \gamma m_k) - \gamma\left[N\frac{\alpha}{\beta} - \frac{\gamma}{\beta}\sum_{C_k \in \mathcal{P}} m_k^2\right]. \tag{5.16f}$$

Remark. As we have noted, a common-property resource game as modeled here is a superior game. However, since there are externalities, the grand coalition may not be stable, and in this case it is not. Denote by $U_i^{\mathcal{G}}$ a payoff for individual i in the grand coalition, for given distributional weights, and denote by U_i^* the payoff to the same individual if the individual deviates as a singleton. Then we have

$$U_i^{\mathcal{G}} = U_i^* - \frac{1}{\beta_i}\left[\gamma_i\sum_{\substack{s=1\\s\neq i}}^{N}\frac{\lambda_s}{\lambda_i}\gamma_s + \frac{1}{2}\left(\sum_{\substack{s=1\\s\neq i}}^{N}\frac{\lambda_s}{\lambda_i}\gamma_s\right)^2\right]. \tag{5.17}$$

Equation (5.17) is expressive of "free-rider" or "holdout" behavior. The implication is that the common-property resource game has a null core, or is characterized by dominance cycles. In application to international fishery agreements, for example, this might account for the difficulty of actually arriving at useful agreements to limit the use of the common property resource, in the absence of any side payments or enforceable penalties for violations. This finding corresponds to a common finding in the literature on coalitions for the sharing of a common property resource.

5.9. The Example of a Cartel

The example of a cartel with differentiated products was introduced in Chapter 3 in Section 3.3, as an instance of an NTU game in partition function form. We have n firms selling products that are imperfect substitutes, and we suppose that $S \subseteq N$ form a cartel.

We recall that this is a superior game; accordingly, we focus attention on the grand coalition, that is, the cartel of all firms in the industry. Each point on the utility possibility frontier for this coalition is characterized by the maximization program corresponding to the following Lagrangean function:

$$\mathcal{L} = \sum_{i=1}^{n} \lambda_i[(p_i - c_i)Q_i - F_i] + \sum_{i=1}^{n} \mu_i \left(\prod_{i=1}^{n} p_j^{\beta_{j,i}} - Q_i \right). \tag{5.18a}$$

Among necessary conditions for the maximum are

$$\frac{\partial \mathcal{L}}{\partial Q_i} = \lambda_i(p_i - c_i) - \mu_i \leq 0 \tag{5.18b}$$

$$\frac{\partial \mathcal{L}}{\partial P_i} = \lambda_i Q_i - \frac{Q_i}{p_i} \sum_{j=1}^{n} \mu_j \beta_{j,i} \leq 0. \tag{5.18c}$$

Assuming interior solutions,

$$\mu_i = \lambda_i(p_i - c_i) \tag{5.18a}$$

$$p_i = \sum_{j=1}^{n} \frac{\lambda_j}{\lambda_i} \beta_{j,i}(p_j - c_j) \tag{5.18b}$$

$$p_i = \frac{\beta_{i,i} c_i}{1 + \beta_{i,i}} - \sum_{\substack{j=1 \\ j \neq i}}^{n} \frac{\lambda_j}{\lambda_i} \frac{\beta_{j,i}}{1 + \beta_{i,i}} (p_j - c_j). \tag{5.18c}$$

Recall that $\beta_{i,j} > 0$ except that $\beta_{ii} < -1$. Therefore, as the second term is positive, $p_i > p_i^*$. (See Eq. (3.21) in Chapter 3.) This corresponds to the common-sense idea that the incentive to an individual firm to deviate from the coalition and go it alone is the freedom to cut its price below the cartel prescribed price for the firm. Second, we see that the cartel would adjust the price of a high-cost firm upward in partial compensation for its high unit costs. Let π_i^* be the profit of firm i in the grand coalition. Thus

$$\pi_i^* = \left[\frac{\beta_{i,i} c_i}{1 + \beta_{i,i}} - \sum_{\substack{j=1 \\ j \neq i}}^{n} \frac{\lambda_j}{\lambda_i} \frac{\beta_{j,i}}{1 + \beta_{i,i}} (p_j - c_j) - c_i \right] Q_i - F_i. \tag{5.18d}$$

Let $\mathcal{P} \in \Pi_N$, $\mathcal{P} \neq \mathcal{G}$. For each $C \in \mathcal{P}$, the price for $i \in C$ will be

$$p_i^{\mathcal{P}} = \frac{\beta_{i,i} c_i}{1 + \beta_{i,i}} - \sum_{\substack{j \in C \\ j \neq i}}^{n} \frac{\lambda_j}{\lambda_i} \frac{\beta_{j,i}}{1 + \beta_{i,i}} (p_j - c_j) \qquad (5.19a)$$

and profit will be

$$\pi_i^{\mathcal{P}} = \left[\frac{\beta_{i,i} c_i}{1 + \beta_{i,i}} - \sum_{\substack{j \in C \\ j \neq i}}^{n} \frac{\lambda_j^{\mathcal{P}}}{\lambda_i^{\mathcal{P}}} \frac{\beta_{j,i}}{1 + \beta_{i,i}} (p_j - c_j) - c_i \right] Q_i - F_i. \qquad (5.19b)$$

Lemma 5.6. $\exists \lambda_j \ni \pi_j > \pi_j^{\mathcal{P}} \; \forall j \in C \in \mathcal{P}$.

Proof. Let $\lambda_j = \lambda_j^{\mathcal{P}} \; \forall j \notin N$. Then

$$p_i^{\mathcal{P}} = p_i + \left(\sum_{j \notin C}^{n} \frac{\lambda_j^{\mathcal{P}}}{\lambda_i^{\mathcal{P}}} \frac{\beta_{j,i}}{1 + \beta_{i,i}} (p_j - c_j) \right) < p_i. \qquad (5.20a)$$

Lower prices for $j \neq i$ implies $Q_i^{\mathcal{P}} < Q_i$ for a given price; thus

$$\pi_i^{\mathcal{P}} < \pi_i + \left(\sum_{j \notin C}^{n} \frac{\lambda_j^{\mathcal{P}}}{\lambda_i^{\mathcal{P}}} \frac{\beta_{j,i}}{1 + \beta_{i,i}} (p_j - c_j) - c_i \right) Q_i < \pi_i. \qquad (5.20b)$$

That is, for an appropriate assignment of utilities (corresponding to an appropriate set of distributive weights) the grand coalition dominates any other partition, consistently with the observation that the game is superior. $\qquad \square$

Let $\mathcal{R} \in \Pi$, $\mathcal{R} \neq \mathcal{G}$, $\mathcal{R} \neq \mathcal{P}$, $C \in \mathcal{R}$. Clearly,

$$\operatorname*{argmax}_{S \in \Xi_C} e_{S, \sigma(\mathcal{R}, S)}(\pi_i^{\mathcal{R}}) = N \in \mathcal{G}. \qquad (5.19c)$$

That is, the grand coalition, considered as a deviation from $\mathcal{R}$, will generate a larger excess than any $S \ni \sigma(\mathcal{R}, S) \neq \mathcal{G}$; since moreover $\sigma(\mathcal{R}, N) = \mathcal{G}$ for any successor function whatever (McCain, 2009, p. 184) we will have

$$\pi_i^{\mathcal{P}} = \pi_i + y_C, \quad y_C < 0. \qquad (5.20c)$$

Noting that

$$y_C = \frac{\sum\limits_{i \in C} [(p_i - c_i)Q_i - F_i - \pi_i]}{|C|} = \frac{Y}{|C|} \tag{5.20d}$$

so that π_i^P increases proportionately with Y. It follows that the coalition will unanimously choose to maximize Y, that is, *unweighted* coalition profits. Therefore, $\lambda_i = \lambda_j = 1$ and Eq. (5.19a) becomes

$$p_i^P = \frac{\beta_{i,i} c_i}{1 + \beta_{i,i}} - \sum_{\substack{j \in C \\ j \neq i}}^{n} \frac{\beta_{j,i}}{1 + \beta_{i,i}} (p_j - c_j). \tag{5.19d}$$

Assuming their prices are determined so that profits correspond to the nucleolus imputation for a given partition, prices will vary according to asymmetries such as differing unit costs. If the conditions of the cartel are symmetric, so that

$$\beta_{j,j} = \gamma \quad \alpha_i = \alpha \qquad Q_i = Q \qquad\qquad \pi_i = \pi$$
$$\beta_{i,j} = \beta \quad c_i = c \quad \text{so} \quad Q_i^P = Q^P \quad \text{and} \quad \pi_i^P = \pi^P$$
$$F_i = F$$

$$p_i^P = \frac{\gamma c}{1 + \gamma} - \sum_{\substack{j \in C \\ j \neq i}}^{n} \frac{\beta}{1 + \gamma} (p - c) = p = \frac{\gamma + (|C| - 1)\beta}{1 + \gamma + (|C| - 1)\beta}. \tag{5.19e}$$

Now, p will be positive and finite at an interior solution if

$$\beta < -\frac{1 + \gamma}{|C| - 1}. \tag{5.21a}$$

In case

$$\beta < -\frac{1 + \gamma}{N - 1} \tag{5.21b}$$

then meaningful monopoly equilibria exist for coalitions of all sizes. In such a case the grand coalition dominates all smaller coalitions, so we may expect that the grand coalition will form and will be stable, with member firms asking equal prices and enjoying equal profits. Otherwise, however, it may be that for large enough $|C|$,

$$\beta > -\frac{1 + \gamma}{|C| - 1}. \tag{5.21c}$$

In such a case there is no meaningful interior solution as profits approach infinity as the price does. This is a familiar conundrum in monopoly theory if the monopoly's own price elasticity is less than one. The novelty is that in this model it may arise despite an own price elasticity less than one because the coalition is large enough that the own price elasticity is not large enough to offset the destabilizing influence of complementarities among the products of the members of the cartel. This difficulty is probably attributable to the simplifying assumption that γ is constant. Plausibly, for a sufficiently high price γ approaches minus infinity. However, a more complex model incorporating that assumption will be beyond the scope of this book.

5.10. Summary and Conclusions

This chapter has proposed a scheme for assigning values to the members of coalitions in a nontransferable utility game in partition function form. The scheme is a generalization of Schmeidler's nucleolus. For a game with strictly convex utility possibility sets for all embedded coalitions, the value imputation is unique and induces a unique set of distributional weights for the members of each embedded coalition, which in turn imply coalition values consistent with the imputations. The imputations take a simple form, as each member of a coalition gets what he would get in the most profitable deviation, plus a bonus (or fee) that is an equal share of the surplus that the coalition generates over those valuations.

The uniqueness of the nucleolus and the corresponding distributional weights distinguishes this value schema from others for NTU games, and, together with applicability to games in partition function form, provide a strong argument for the nucleolus as the most broadly applicable of value imputations in coalitional games.

Chapter 6

A Core Imputation with Variable Bargaining Power

Among the various solution concepts for cooperative games, the core has been widely studied, in part, no doubt because it reflects an appealing concept of stability against group deviations. However, the core may be null or may comprise a continuum of imputations among the members of a coalition. The latter possibility poses the problem of selecting an imputation in the core. Similar problems arise in models of hiring in the tradition of the Nobel Laureates Diamond (1982), Mortensen (1982), and Pissarides (1985). See, e.g., Shimer (2005), Hall (2005a), and Hall and Milgrom (2008). It is often proposed that this non-uniqueness can be resolved by bargaining. Such a problem of indeterminacy has also arisen recently in papers by Brandenburger and Stuart (2007) and Chatain and Zemsky (2007), who explicitly use the core of a cooperative game in their model. This also is a matter of bargaining.

For two-person games, Nash bargaining theory is probably the most widely used approach, but despite the early work of Harsanyi (e.g., 1963) there is no widely accepted generalization of the Nash bargaining model to $N > 2$ participants. Generalizations by Roth (1979) and Svejnar (1986) apply only to the special case of "the bargaining problem," i.e., to games in which no coalition other than the grand coalition adds value to the fine partition, and not to N-person games in general. The Shapley (1953) value can be calculated for any game in coalition function form, and assigns a unique imputation, but may be outside the core. Schmeidler's nucleolus (1969; see also Aumann and Dreze, 1974; McCain, 2009) can also be calculated for any game in coalition function form, is unique, and is an element of the core when the core is not null. Thus the nucleolus might provide a resolution to the core imputation problem. In a certain sense, the nucleolus assumes equal bargaining power for all groups of participants, however. In theory, this may be sound: if there are differences in bargaining power, we would like to make the causes of the difference part of our model. From a pragmatic viewpoint, however, that may not be possible, and there is some evidence (Svejnar, 1986) that a bargaining theory fits the data significantly better if it allows for exogenous differences in bargaining power.

In two-person bargaining theory, Frederik Zeuthen (1930) originated the model in which the bargain maximizes the product of the gains of the two bargainers, and the same result arose from Nash's (1950, 1953) bargaining theory, and the extensions of Zeuthen–Nash bargaining theory to unequal bargaining power can also be expressed as maximization formulae. Schmeidler's nucleolus also rests on an optimization hypothesis. He defines the *excess* for a potential deviation from a coalition structure as the difference between the coalition's value as a separate coalition and the sum of their imputations in the current coalition. The excess for the deviation then can be thought of as a measure of their relative dissatisfaction with the current imputation. The nucleolus is defined as the imputation that minimizes the maximum excess. As, however, Forgo *et al.* (1999) observe, the min–max function is not the only possibility. A nucleolus-like solution could also be obtained by minimizing any increasing aggregation function of the excesses in the game. With a normative concept of fairness in mind, they propose a least-squares nucleolus that minimizes the sum of the squared differences of the excesses of the different agents in the game.

For the different and non-normative purposes of this chapter, an optimization model of bargaining will be adopted that reduces to the Zeuthen–Nash model in the case of two persons and equal bargaining power. For Sections 6.1–6.4, the game will be a TU game in coalition function form, while Section 6.5 extends the solution to NTU games in partition function form. In this way the chapter simultaneously (1) extends Zeuthen–Nash bargaining theory to N-person games in coalition function form, with varying bargaining power, and (2) extends the nucleolus concept to a case of varying bargaining power with substitutability among the excesses of different coalitions. Unlike any of these predecessors, the model proposed in this chapter can include cases in which agents with little or no individual bargaining power can gain or increase their bargaining power by leaguing together, as in a cartel or labor union.

6.1. Bargaining Power Games

A Bargaining Power Game (BP game) in coalition function form comprises a group of agents $N = \{1, 2, \ldots, n\}$, a coalition value function v, and a *power function* $\phi_S \forall S \subseteq N$[1] with $\forall S$, $\phi_S \geq 0$. Transferable utility is assumed. Now, let $\mathcal{P}$ be a coalition structure (i.e., a partition of N into disjoint coalitions), $C \subseteq N$, $S \subseteq N$, $C \subseteq \mathcal{P}$, $S \notin \mathcal{P}$, (thus by definition S is a deviation) $\mathbf{x}$ an imputation; then $x_C = \sum_{i \in C} x_i$, $x_S = \sum_{i \in S} x_i$.

We might follow Schmeidler (and Aumann and Dreze, 1974) in defining the *excess* for S, given $\mathbf{x}$ and a given coalition structure $\mathcal{P}$, as $e_{\mathcal{P},S}(\mathbf{x}) = v(S) - x_S$. For present purposes, however, it will be more convenient to reverse the sign and define the *gain* of S as $g(\mathcal{P}, S, \mathbf{x}) = x_S - v(S)$. If the gain is negative for any S, then $\mathbf{x}$ is not in the core for $\mathcal{P}$.

Let $\Xi_C = \{S \subseteq N | S \notin \mathcal{P}, S \cap C \neq \varnothing\}$ and let $\Xi_i = \{S \in \Xi_C, i \in S \cap C\}$ as in the previous chapter. Now, we will define the *N-S nucleolus* for coalition structure $\mathcal{P}$ as

$$\mathbf{y} = \arg\max_{\mathbf{x}} \sum_{S \in \Xi_C} \phi_S \ln g(\mathcal{P}, S, \mathbf{x}) \tag{6.1}$$

subject to

$$x_C = v(C) \quad \forall C \in \mathcal{P} \tag{6.1a}$$

and

$$g(\mathcal{P}, S, \mathbf{x}) \geq 0 \quad \forall S \in \Xi_C. \tag{6.1b}$$

Constraints (6.1a) follow Aumann and Dreze and McCain (2009) in assuming that each coalition "sits on its own bottom," distributing among its members no more than its own value, i.e., there are no side payments across coalitions. The constraints are equality constraints because we suppose that the imputation will be Pareto-optimal within each coalition, and, in a TU game, this means that the value of the coalition must be exhausted. Constraints (6.1b) are the individual and group rationality

---

[1]Note that formally we are assigning a bargaining power to the grand coalition among others, and that may seem odd, since bargaining power is exerted against other groups. If, however, we consider imputations for a coalition structure other than the grand coalition, the grand coalition is a potential deviation, and we need an indication of its bargaining power *as a deviation*.

constraints. If there is no feasible solution to this problem then the core is null for coalition structure $\mathcal{P}$. If there is a unique feasible imputation such that the solution is constraint-determined, then the core for coalition structure $\mathcal{P}$ comprises that unique imputation. As our concern here is with a core imputation we may neglect these cases.

Note that (6.1) is equivalent to

$$\mathbf{y} = \arg\max_{\mathbf{x}} \prod_{S \in \Xi_C} g(\mathcal{P}, S, \mathbf{x})^{\phi_S}. \tag{6.1*}$$

Moreover, for $i \in C \backslash S$, $j \in C \cap S$, we have

$$\frac{\partial g_S}{\partial x_j} = 1, \quad \frac{\partial g_S}{\partial x_i} = 0,$$

$$\frac{\partial x_C}{\partial x_j} = 1, \quad \frac{\partial x_C}{\partial x_i} = 1. \tag{6.1c}$$

The Lagrangean function is

$$\mathcal{L} = \sum_{S \in \Xi_C} \phi_S \ln g(\mathcal{P}, S, \mathbf{x}) + \mu_C \left(v(C) - \sum_{i \in C} x_i \right) + \sum_{S \in \Xi_C} \mu_S g(\mathcal{P}, S, \mathbf{x}). \tag{6.2}$$

Note that the μ_C may take negative values while the μ_S are non-negative.

Using (6.1c), we then have the necessary conditions for the maximum as

$$\frac{\partial \mathcal{L}}{\partial x_i} = \sum_{S \in \Xi_i} \frac{\phi_S}{g(\mathcal{P}, S, \mathbf{x})} - \mu_C + \sum_{S \in \Xi_i} \mu_S \le 0. \tag{6.3a}$$

Note that $\{i\} \in \Xi_{\{i\}}$ by identity. Now, consider $C \in \mathcal{P}$, and let m_C be the cardinality of C. Thus we have

$$\sum_{S \in \Xi_C} \frac{\phi_S}{g(\mathcal{P}, S, \mathbf{x})} \le \mu_C - \sum_{S \in \Xi_C} \mu_S \tag{6.3b}$$

that is,

$$\mu_C \ge \sum_{S \in \Xi_C} \left(\frac{\phi_S}{g(\mathcal{P}, S, \mathbf{x})} - \mu_S \right) \tag{6.3c}$$

This will hold as in equality if there is an interior solution for any $i \in C$, as there will be if the coalition produces any surplus value whatever. That is, the marginal value of payments to coalition C is no less than the average value of its members' bargaining powers, adjusted for the satisfaction of each potential deviation with its payoffs at the current imputation (the deviation's gain) and adjusted for the marginal value of payments required to prevent members of the coalition from joining in deviations that would otherwise be profitable. This will be an equality in case we have interior solutions for all members of C. These conditions will apply for each $C \in \mathcal{P}$.

Let $T \subseteq N$, $T \notin \mathcal{P}$ i.e., T is a potential deviation from $\mathcal{P}$. Let m_T be the cardinality of T. For any Agent i denote by C_i the coalition in $\mathcal{P}$ to which i belongs. Using (6.3a), we have

$$\sum_{i \in T} \sum_{S \in \Xi_i} \left[\frac{\phi_S}{g(\mathcal{P}, S, \mathbf{x})} + \mu_S \right] - \sum_{i \in T} \mu_{C_i} \leq 0. \qquad (6.4a)$$

That is, from (6.3b),

$$\sum_{i \in T} \left[\frac{\phi_T}{g(\mathcal{P}, T, \mathbf{x})} + \mu_T \right] + \sum_{i \in T} \sum_{\substack{S \in \Xi_i \\ S \neq T}} \left[\frac{\phi_S}{g(\mathcal{P}, S, \mathbf{x})} + \mu_S \right] \leq \sum_{i \in T} \mu_{C_i}. \qquad (6.4b)$$

Then

$$\frac{m_T \phi_T}{g(\mathcal{P}, T, \mathbf{x})} \leq \sum_{i \in T} \mu_{C_i} - m_T \mu_T - \sum_{i \in T} \sum_{\substack{S \in \Xi_i \\ S \neq T}} \left[\frac{\phi_S}{g(\mathcal{P}, S, \mathbf{x})} + \mu_S \right] \qquad (6.4c)$$

$$g(\mathcal{P}, T, \mathbf{x}) \geq \phi_T \left[\frac{m_T}{\displaystyle\sum_{i \in T} \mu_{C_i} - m\mu_T - \sum_{i \in T} \sum_{\substack{S \in \Xi_i \\ S \neq T}} \left[\frac{\phi_S}{g(\mathcal{P}, S, \mathbf{x})} + \mu_S \right]} \right]. \qquad (6.4d)$$

Thus the gain for a deviation T is no less than a proportion of the bargaining power of T. This will be an equality if we have an interior solution for any member of T. Note that if ϕ_T is positive then, by the comparative slackness conditions, $\mu_T = 0$, so the term (6.4d) becomes

$$g(\mathcal{P}, T, \mathbf{x}) \geq \phi_T \left[\frac{m_T}{\displaystyle\sum_{i \in T} \mu_{C_i} - \sum_{i \in T} \sum_{\substack{S \in \Xi_i \\ S \neq T}} \left[\frac{\phi_S}{g(\mathcal{P}, S, \mathbf{x})} + \mu_S \right]} \right]. \qquad (6.4e)$$

The gain for a deviation T is increased as a proportion of the bargaining power of T by an increase in the adjusted bargaining power of other deviations that the members of the deviation might participate in, $\frac{\phi_S}{g(\mathcal{P},S,\mathbf{x})}$, or the marginal value of payments to those deviations to stabilize the coalition structure against their deviation, μ_S, and decreased by the marginal values of the coalitions in which the members of the deviation participate, $\sum_{i \in T} \mu_{C_i}$. Let $T = \{i\}$. Then (6.4e) simplifies to

$$g(\mathcal{P},\{i\},\mathbf{x}) \geq \phi_{\{i\}} \left[\frac{1}{\mu_{C_i} - \sum_{\substack{S \in \Xi_i \\ S \neq \{i\}}} \left[\frac{\phi_S}{g(\mathcal{P},S,\mathbf{x})} + \mu_S \right]} \right] \tag{6.4f}$$

so that the gain to each individual is no less than proportional to the *individual's* bargaining power as a singleton coalition, and this is an equality in the case of an interior solution for i.

Suppose that $i \in C$, $j \in C$, $C \in \mathcal{P}$, and suppose we have an interior solution for both i and j. Then

$$\frac{g(\mathcal{P},\{i\},\mathbf{x})}{g(\mathcal{P},\{j\},\mathbf{x})} = \frac{\phi_{\{i\}}}{\phi_{\{j\}}} \left[\frac{\mu_{C_j} - \sum_{\substack{S \in \Xi_j \\ S \neq \{j\}}} \left[\frac{\phi_S}{g(\mathcal{P},S,\mathbf{x})} + \mu_S \right]}{\mu_{C_i} - \sum_{\substack{S \in \Xi_i \\ S \neq \{i\}}} \left[\frac{\phi_S}{g(\mathcal{P},S,\mathbf{x})} + \mu_S \right]} \right]. \tag{6.4g}$$

That is, the gain of coalition C will not in general be divided between its members simply in proportion to their given bargaining powers, but these bargaining powers will have to be adjusted for the values of the different deviations they might participate in, including, for i (for example) all deviations that include i but exclude j, and conversely.

If, however, we impose $\phi_S > 0 \ \forall S \notin \mathcal{P}$, i.e., every deviation has some positive bargaining power, then $\mu_S = 0$ for all deviations and Eq. (6.3c) becomes

$$\mu_C = \frac{1}{m_C} \sum_{i \in C} \sum_{S \in \Xi_C} \frac{\phi_S}{g(\mathcal{P},S,\mathbf{x})} \tag{6.3d}$$

Expression (6.4e) becomes

$$g(\mathcal{P}, T, \mathbf{x}) \geq \phi_T \left[\frac{m_T}{\displaystyle\sum_{i \in T} \frac{1}{m_{C_i}} \sum_{j \in C_i} \sum_{S \in \Xi_{C_j}} \frac{\phi_S}{g(\mathcal{P}, S, \mathbf{x})} - \sum_{i \in T} \sum_{\substack{S \in \Xi_i \\ S \neq T}} \frac{\phi_S}{g(\mathcal{P}, S, \mathbf{x})}} \right]$$

(6.4h)

and (6.4f) becomes

$$g(\mathcal{P}, \{i\}, \mathbf{x}) \geq \phi_{\{i\}} \left[\frac{1}{\displaystyle\sum_{S \in \Xi_i} \frac{\phi_S}{g(\mathcal{P}, S, \mathbf{x})} - \sum_{\substack{S \in \Xi_i \\ S \neq \{i\}}} \frac{\phi_S}{g(\mathcal{P}, S, \mathbf{x})}} \right].$$

(6.4j)

This model allows groups as well as individuals (singleton coalitions) to have bargaining power. An alternative, commonsense view would attribute bargaining power only to individuals. (In two-person bargaining, the topic of most bargaining theory, the dichotomy does not arise.) If this assumption is imposed, then (6.3b) is simplified to (6.3d), (6.4e) reduces to the group rationality constraints for coalitions with two or more members, and (6.4f) is reduced to (6.4k) and (6.4g) to (6.4m). In some applications this will be an appropriate simplification.

$$\mu_C \geq \frac{1}{m_C} \sum_{i \in C} \left(\frac{\phi_{\{i\}}}{g(\mathcal{P}, \{i\}, \mathbf{x})} + \mu_{\{i\}} \right)$$

(6.3d)

$$g(\mathcal{P}, \{i\}, \mathbf{x}) \geq \frac{\phi_{\{i\}}}{\mu_{C_i}}$$

(6.4k)

$$\frac{g(\mathcal{P}, \{i\}, \mathbf{x})}{g(\mathcal{P}, \{j\}, \mathbf{x})} = \frac{\phi_{\{i\}}}{\phi_{\{j\}}}.$$

(6.4m)

In this case, from (6.4m), the benefits from a coalition will be divided among the members in proportion to their bargaining powers, assuming that all have positive bargaining power so that there will be an interior solution for each of them. Thus, the limitation of bargaining power to singleton deviations is a powerful simplifying assumption, particularly if we assume that each singleton coalition has positive bargaining power.

A contrary intuition suggests that individuals may gain bargaining power by leaguing together even if they lack bargaining power as individuals. This, after all, is the rationale of every labor union and cartel. Alternatively, a group may derive bargaining power from a credible threat to exclude some members of a larger coalition. In general, bargaining powers may interact, so that the bargaining power of a group may be more (or less) than the bargaining power of its members. The theory suggested here is general and applicable in either case.

6.2. Examples

Consider the following examples[2]:

(a) Game 6.1 is summarized in Table 6.1. The coalition structure of interest, C, will be the grand coalition. The assumptions with respect to coalition values are (1) that Agent d brings to the coalition some resource or opportunity that complements the resources or opportunities brought by a, b, and/or c, so that (apart from singletons) there is no value without d present in the coalition, but the value is proportionate to the number of a, b, c in the coalition. With respect to bargaining power, it is assumed that no individual has positive bargaining power as a singleton coalition, but a deviation of two or more agents may have positive bargaining power, and a coalition of a, b, c has bargaining power no less than any other coalition. The solution is as shown in Table 6.2. For this problem λ_C is 1.38. All other Lagrange multipliers are zero.

(b) Game 6.2 differs from Game 6.1. only in that $v(\{a\}) = v(\{b\}) = v(\{c\}) = 0$. The solution is identical. Since the solution is determined by the relative bargaining powers of coalitions of more than one member, their values as singleton coalitions (that is, in economic terms, their opportunity costs) may vary widely without influencing the solution.

(c) Game 6.3 differs from Game 6.1 in that $v(\{a\}) = v(\{b\}) = v(\{c\}) = 4$. The solution is in Table 6.3. The nonzero Lagrange multipliers are

---

[2]Solutions for these examples have been computed using solver software supplied with the Microsoft Excel spreadsheet.

Table 6.1. Game 6.1: A
four-person game.

Coalitions	v	ϕ
$\{a, b, c, d\}$	20	5
$\{a, b, c\}$	0	5
$\{a, b, d\}$	13.2	1
$\{a, c, d\}$	13.2	1
$\{b, c, d\}$	13.2	1
$\{a, b\}$	0	0.5
$\{a, c\}$	0	0.5
$\{a, d\}$	6.6	1
$\{c, d\}$	6.6	1
$\{b, d\}$	6.6	1
$\{b, c\}$	0	0.5
$\{a\}$	2	0
$\{b\}$	2	0
$\{c\}$	2	0
$\{d\}$	0	0

Table 6.2. The solution
to Game 6.1.

a	3.51
b	3.51
c	3.51
d	9.47

Table 6.3. The solution to
Game 6.3.

a	4.00
b	4.00
c	4.00
d	8.00

shown in Table 6.4. From this example, we see that the imputation will
not be independent of the singleton values (opportunity costs) in every
case, but instead the singleton values may determine the imputation if
they are large enough.

(d) Game 6.4 differs from Game 6.1 only in that $\phi_{\{b,c\}}$ is reduced to zero.
The solution is shown in Table 6.5. We see in this case that even where
the coalition function is symmetrical for a group of players, asymmetries

Table 6.4.   **Lagrange multipliers
for Game 6.3 Solution.**

λ_C	1.63
$\lambda_{\{a\}}$	0.19
$\lambda_{\{b\}}$	0.19
$\lambda_{\{c\}}$	0.19

Table 6.5.   **The solution
for Game 6.4.**

a	3.79
b	3.14
c	3.14
d	9.93

Table 6.6.   **A solution
for Game 6.5.**

a	6.00
b	2.00
c	2.00
d	10.00

of the bargaining powers of subsets of them will induce asymmetries in the solution.

(e) Game 6.5 differs from Game 6.1 in the bargaining powers, and assumes that $\phi_{\{a,b,c\}} = \phi_{\{d\}} = \phi_{\{a,b,c,d\}} = 5$ and all other bargaining powers are zero. A computed solution for this game is shown in Table 6.6. However, the solution is not unique and the conditions for the problem are satisfied for any imputation that divides equally between $\{a,b,c\}$ and $\{d\}$. The problem is underdetermined, and this may be the case if the bargaining powers of many coalitions are zero.

(f) Game 6.6 differs from Game 6.1 only in that the zero bargaining powers of the singleton coalitions are replaced by bargaining powers of 0.01. The solution is as shown in Table 6.7. Comparing Table 6.7 with Table 6.2, we see that there are differences only in the second decimal. This suggests that the assumption that all bargaining powers are positive may be innocuous, in that, as we expect, the solution to a model with very small bargaining powers substituted for zero

Table 6.7. The solution
of Game 6.6.

a	3.52
b	3.52
c	3.52
d	9.43

Table 6.8. The solution to
Game 6.7.

a	3.35
b	3.35
c	3.35
d	9.95

Table 6.9. The solution to
Game 6.8.

a	3.75
b	3.75
c	3.75
d	8.75

bargaining powers is likely to approximate the solution with zero
bargaining powers.

(g) Game 6.7 then differs from the Game 6.5 only in that all zero bargaining
powers are replaced by bargaining powers of 0.01. The solution to this
game is shown as Table 6.8. We see that the solution is now determinate,
in a way that reflects the symmetry of a, b, and c, but the asymmetry
of bargaining power between $\{a, b, c\}$ and $\{d\}$ is slightly reduced.

(h) Game 6.8 retains the same coalition function as Game 6.1 but assumes
the bargaining powers for all coalitions with two or more members are
zero and $\phi_{\{a\}} = \phi_{\{b\}} = \phi_{\{c\}} = 1$ while $\phi_{\{d\}} = 5$. The solution is shown
in Table 6.9.

(i) Game 6.9 is similar to Game 6.8, except that as in Game 6.2,
$v(\{a\}) = v(\{b\}) = v(\{c\}) = 0$. The solution is shown in Table 6.10.

Tables 6.9 and 6.10 could be computed by hand using Eq. (6.4.1), but
they illustrate a key point with respect to the assumption that only individuals have bargaining power. The point is that bargaining is reduced

Table 6.10. The solution
to Game 6.9.

a	2.5
b	2.5
c	2.5
d	12.5

in effect to a set of two-way bargains, so that the fact that d has five times the bargaining power as any of a, b, c means that his gain, at an interior solution, is five times the gain of any one of the others. If the others were more numerous the total gain of those with less individual bargaining power could nevertheless be greater. And the gain of the powerful bargainer would decline accordingly until the powerful bargainer's individual rationality constraint becomes binding. Conversely, we see that the payoff to each of a, b, and c decreases along with their opportunity costs, but while their opportunity costs decrease by 2, their payoffs decline only by 1.25. This reflects the fact that the decline in their opportunity costs increases the total surplus generated by the grand coalition, and the increase in total surplus is distributed across all agents in proportion to their bargaining power.

6.3.   The Relation of this Model to Some Predecessors

In effect, if not explicitly, Nash and Schmeidler assumed equal bargaining powers for all deviations. Before addressing the case of equal bargaining power, we note that the model of Section 6.1 reduces to the Roth–Svejnar model in the case of two-person games.

We suppose that there are just two agents $i = 1, 2$, $\mathcal{P} = \mathcal{G} = \{N\}$ where $\mathcal{G}$ is the grand coalition and $\{S|S \notin \mathcal{P}\} = \{\{1\}, \{2\}\}$. Their bargaining powers are $\alpha > 0$ and $\beta > 0$. Without loss of generality $\beta = 1 - \alpha$. Thus $v(\{1\}) = v_1$ is the threat outcome for bargainer 1 and $v(\{2\}) = v_2$ is the threat outcome for bargainer 2. Expression 1* now is

$$\{x_1, x_2\} = \arg \max(x_1 - v_1)^\alpha (x_2 - v_2)^\beta \qquad (6.5a)$$

which is the Roth–Svejnar extension of the Zeuthen–Nash theory.

From (6.3b),

$$\mu_\mathcal{G} = \frac{1}{2}\left[\frac{\alpha}{x_1 - v_1} + \frac{\beta}{x_2 - v_2}\right], \quad \mu_{\{1\}} = 0, \quad \mu_{\{2\}} = 0. \qquad (6.5b)$$

Noting that as

$$\Xi_i/\{T\} = \varnothing, \quad \sum_{\substack{S \in \Xi_i \\ S \notin \mathcal{P}}} \frac{\phi_S}{g(\mathcal{P}, S, \mathbf{x})}$$

is a summation over a null set and is zero, from (6.4h) we have

$$x_1 - v_1 = \alpha \left[\frac{1}{\frac{1}{2}\left(\frac{\alpha}{x_1 - v_1} + \frac{\beta}{x_2 - v_2}\right)} \right] \tag{6.5c}$$

$$(x_1 - v_1)\frac{1}{2}\left(\frac{\alpha}{x_1 - v_1} + \frac{\beta}{x_2 - v_2}\right) = \alpha \tag{6.5d}$$

$$\alpha + \beta \left[\frac{x_1 - v_1}{x_2 - v_2}\right] = 2\alpha \tag{6.5e}$$

$$\frac{x_1 - v_1}{x_2 - v_2} = \frac{\alpha}{\beta}. \tag{6.5f}$$

That is, for a TU game, the benefits of the coalition are divided in proportion as the bargaining powers of the two agents.

To obtain the Zeuthen–Nash result, we add the restriction that the two bargainers have equal bargaining power, i.e., $\alpha = \beta$, and without loss of generality $\alpha = \beta = 1$. Thus (6.5a) becomes

$$\{x_1, x_2\} = \arg\max(x_1 - v_1)(x_2 - v_2) \tag{6.6a}$$

and (6.5f) yields

$$x_1 - v_1 = x_2 - v_2 \tag{6.6b}$$

since, for a TU game, the benefits of a coalition are always divided equally in Zeuthen–Nash bargaining theory.

The model of this chapter does *not* reduce to Schmeidler's nucleolus in any special case, and thus it cannot be considered a generalization of Schmeidler's analysis in quite the same sense as it generalizes those of Nash, Zeuthen, Roth and Svejnar. This is the case because Eq. (6.1*), the equivalent maximization, is Cobb–Douglas and as such assumes that the elasticity of substitutions among the gains of different deviations is always one. By contrast, Schmeidler's min–max function assumes an elasticity of substitution of zero. We might, however, consider both bargaining theories as instances of a still larger class that would allow any constant elasticity of substitution. We would then identify Schmeidler's nucleolus as the special case of equal bargaining powers and zero elasticities of substitution and the N-S nucleolus as the special case in which elasticities of substitution are one.

6.4. Matching Games and Bargaining Power

In work recently honored by the Nobel Memorial Prize in Economics, Mortensen and Pissarides and Diamond (MPD) explain the persistence and dynamic properties of unemployment by the plausible assumption that there are significant costs of matching job seekers with potential employers.[3] Once the match is formed, the employer and employee enter into a cooperative coalition.[4] In the MPD framework, this coalition generates a surplus, so that the division of the surplus is not determined by competition. So long as the wage is between the productivity of the marginal worker and the productivity minus the (appropriately discounted) expected cost of recruiting, the employer has no incentive either to hire or to reduce the firm's work force by dismissal. If the expected period of unemployment is sufficiently high, neither has a representative worker any incentive to leave the employment relation. Thus, the coalition of the employer and a group of employees is a stable one.

The fact that employment under matching costs generates a surplus presents a bargaining problem. One approach has been to assume that the surplus is evenly split between the employee and the employer[5] (Mortensen

---

[3]In the chapters to follow, we will return to matching games and will consider them in terms of Biform Games. In that model, the decision to enter labor markets and to apply for or extend an offer of employment are noncooperative ones; but the ongoing employment relationship is itself a cooperative one. Here we consider matching games mainly for the bargaining problem they present.

[4]This is far from a new idea. See, e.g., Aoki (1980) and McCain (1980).

[5]Mortonson and Pissarides (1994) note the bargaining problem implied by the matching cost model, and is the source of much of the subsequent literature. Shimer (2005) criticized Mortonson and Pissarides and related literature in that these models are inconsistent with the observed "stickiness" of wages and volatility of unemployment. Hall's papers (2005a, 2005b) and Hall and Milgrom (2008) have particularly influenced the discussion in this chapter.

and Pissarides, 1994). This has been interpreted as a Nash bargain[6] and is consistent with Nash' theory when the determination of the wage is modeled as a two-person bargain over a fixed coalition value. However, the results do not seem to agree with the empirical evidence. (See esp. Shimer, 2005) The difficulty is that the Nash bargain (in this sense) depends strongly on the "threat outcomes" of the bargainers. Thus, when unemployment rises, the threat outcome for employees deteriorates with it, and so the wage bargain declines. This leads to the prediction that shocks will result in relatively wide fluctuations in wages but quite stable rates of unemployment, which is roughly the opposite of what we observe.

This chapter has argued, however, that the bargaining power of a group may be quite different from the bargaining power of the individuals who make up the group. In principle, the overall wage level for an enterprise is a bargain between the owners of the enterprise and the employees of the enterprise as a group. As we have seen in Games 6.1 and 6.2, the payoffs to a group of similar agents may be entirely independent of the values that the individuals can realize if they withdraw from the coalition as singleton coalitions. If we interpret a, b, and c in those examples as employees, and d as an employer, this is to say that within some range the wage bargain would be independent of the rate of unemployment and the alternative opportunities of employees.[7] In the examples, no individual (including the

---

[6]This is not explicitly assumed by Mortonson and Pissarides. However, they write (p. 398) "Wages are the outcome of a bilateral bargain that takes place when unmatched jobs and workers meet and *is revised continually in the face of productivity shocks.*" They add (p. 399) "wages are chosen so as to share *at all times* the surplus from a job match in fixed proportions. The worker's share is β." (Emphasis added in both cases.) They then specify (p. 411) that $\beta = \frac{1}{2}$.

[7]Hall (2005a) analyzes the bargaining in terms of the Nash (1953) Demand Game, a noncooperative bargaining model which, as he notes, has a continuum of solutions. This is consistent with a self-enforcing convention as the determinant of the realized Nash equilibrium, and Hall (2005b) suggests that the wage may indeed be determined by convention. However, Hall and Milgrom (2008) substitute an alternating-offer bargaining theory in which the time delay in the bargaining process itself is the key determinant of the bargain. This approach to bargaining theory is formalized in game theoretic terms by Rubinstein (1982) but seems closer to that of Cross (1975) than to Nash or Zeuthen. Thus, in Hall and Milgrom, the worker's bargaining power is not derived from a threat to break off bargaining and return to search but the threat to delay the bargaining by making a counteroffer. As usual, the threat is not carried out in equilibrium. The credibility and cost of the threat of delay is largely independent of the unemployment rate, so the wage determined in this way also is. This is, however, a theory of the wage bargain in the first period of employment, not a theory of the overall wage level, which will be dominated by the wages paid to employees in long-term employment relationships

employer) has any bargaining power *as an individual*, but the wage bargain is determined by the interplay of the bargaining power of groups of a, b, and c, including in particular $\{a, b, c\}$, in their threats to withdraw from the coalition as a group, *vis-à-vis* the bargaining power of Agent d via Agent d's threats to withdraw along with one or two of a, b, and c, that is, the dismissal threat to exclude some of the agents a, b, and c from a coalition that would then still have a value proportionate to the number who remain.

One obvious way in which $\{a, b, c\}$ as a group might gain bargaining power that is unavailable to any of them as individuals is through the threat of a strike. Of course, a strike is not a credible threat in a noncooperative sense, since it makes the employees worse off; but among the things that distinguish a cooperative from a noncooperative game is that in a cooperative game, threats that make the threatener worse off may nevertheless influence the solution. In that perspective, one factor in explaining the bargaining power indices ϕ could be the relative plausibility of the strike threats and of any retaliatory threats that might be made. If a relatively small subset of the employees of a firm were to try to strike, the threat of retaliatory dismissal would be plausible (although, given hiring costs for their replacements, it might make the employer worse off) and their own threat is correspondingly less plausible. By contrast, if the employees threaten a strike on a unified basis — or very nearly so — then their own threat is more plausible and the retaliatory threat of the dismissal and replacement of the whole group is less plausible (though it has sometimes been done). While only a small proportion of U.S. employees are represented by labor unions, strikes can occur in the absence of unions, and the threat of unionization may also contribute something to the bargaining power of the work group as a whole.

It seems likely, however, that withdrawal of effort is a threat that lends more to the bargaining power of the working group as a unified group. This threat could be highly credible. First, it may be a move toward a noncooperative equilibrium, thus requiring no enforcement. Indeed, it is the maintenance of higher effort level that would require enforcement, in

---

simply because they are more numerous. These wages, too, will be "revised continually in the face of productivity shocks." Thus, this chapter diagnoses the problem differently, and argues that the problem arises precisely from the assumption that "wages are the outcome of a bilateral bargain that takes place when unmatched jobs and workers meet" when they are instead the outcome of a more complex bargain among the participants in the ongoing employment relation.

the noncooperative perspective. Second, the retaliatory threat of dismissal is less credible where all, or a large proportion, of employees withdraw their effort than would be the case if a single employee were to shirk. This will be especially true if the cost of hiring rises more than in proportion to the rate of hiring. It is true that this threat poses a problem of coordination among the employees who withdraw their effort, but this coordination may occur spontaneously and even unconsciously via a "grapevine." Alternatively, an action by the employer, such as a wage cut or increase in hours, could provide the signal needed for coordination. All of these are *possibilities*, and there may be others. The key points to be made are that (1) the bargaining power of the work group as a whole will have a powerful impact on the wage level and (2) the bargaining power of the work group as a unified group may be determined by permanent features of the organization that are largely independent of the opportunity cost or threat outcome for an individual employee. These *possibilities* do seem to be consistent with the observed *facts* that have been expressed as "the limited influence of unemployment on the wage bargain" (Hall and Milgrom, 2008).

6.5. Bargaining Power NTU Games in Partition Function Form

We now extend the results of this chapter to the case of NTU games in partition function form, relying on conventions developed in Chapters 3 and 5. Accordingly, we have a set of agents $N = \{1, 2, \ldots, n\}$ and a power function ϕ_i as before, but in place of the coalition value function we have

$$f : \{(\mathcal{P}, C) | C \in \mathcal{P}\} \to \{f_{\mathcal{P}, C}\} \qquad (6.7)$$

where $\mathcal{P}$ is a partition of N and $f_{\mathcal{P}, C}$ is a constraint function, which for the purposes of this section we will assume to be strictly convex.

Now let $S \notin \mathcal{P}$ a deviation from $\mathcal{P}$. Let $\mathcal{R} = \sigma(\mathcal{P}, S)$. An imputation $\{U_i^{\mathcal{P}}\}$ will be admissible for $\mathcal{P}$ if, $\forall C \in \mathcal{P}$, $f_{\mathcal{P}, C}(\mathbf{U}_C) \leq 0$ where $\mathbf{U}_C = \{U_i^{\mathcal{P}}\}_{i \in C}$. In Chapter 5, Section 5.1, the excess (available for the members of S through their deviation to form S) is defined as

$$e(\mathcal{P}, S, U_i) = \min_{i \in S}(U_i^{\mathcal{R}} - U_i^{\mathcal{P}}) \qquad (6.8)$$

where $U_i^{\mathcal{R}}$ is the imputation expected in $\mathcal{R}$. As has been done previously in this chapter, we will instead define a gain (to the members of S from remaining in $\mathcal{P}$ rather than deviating to S).

The informal reasoning underlying (6.8) was that a coalitional compensation committee would want to reduce the discontent of the most discontented group. In a model with well-defined and differentiated bargaining power, we would expect instead that the compensation committee would distribute the coalition's surplus partly according to bargaining power. Accordingly, corresponding to a realized Pareto-optimal imputation U_C for coalition C there will be a set of distributional weights λ_i such that

$$U_C = \arg \max \sum_{i \in C} \lambda_i U_i \qquad (6.9)$$

for the coalition. The distributional weights λ_i will be determined, along with the U_i, by bargaining power. This maximum might be interpreted as the value of the coalition C. Accordingly, the gain for a deviation S from $\mathcal{P}$ should also be measured in the units defined by λ_i.

Our purpose is to obtain a core assignment solution, so it will be necessary first to recall and extend the characterization of the core for a game such as this. As discussed at Chapter 5, Section 5.4, for a deviation S with $\mathcal{R} = \sigma(\mathcal{P}, S)$, the rationality constraint will be

$$f_{\mathcal{R},S}(U_S^{\mathcal{P}}) \geq 0. \qquad (6.10)$$

Taking λ_i for coalitions in $\mathcal{P}$ as given, for the present, we might define the value of a deviation $S \notin \mathcal{P}$ as

$$V_{\mathcal{P}}(S) = \max_{U_i^{\mathcal{R}}} \lambda_i U_i^{\mathcal{R}}$$

subject to

$$f_{\mathcal{R},S}(U_i^{\mathcal{R}}) \leq 0. \qquad (6.11a)$$

This represents the best offer that the deviation can make to its members, measured in units of the distributional weights of the embedded coalition $\mathcal{P}$, C. Now, the potential members of S will themselves expect payments that maximize the gain from S relative to further deviations from $\mathcal{R}$. However, when we take this expectation into account, we would be characterizing an extended core (Chapter 5, Section 5.6) rather than a conventional core imputation.

As in the Chapter 5, Section 5.2, consider a constructed *noncooperative* game in which the players are embedded coalitions, the strategies are the utility assignments, and the payoff for $\mathcal{P}$, S is

$$g(\mathcal{P}, S, \mathbf{U}) = \sum_{i \in S} \lambda_i (U_i^{\mathcal{P}} - U_i^{\mathcal{R}}) \tag{6.11b}$$

where $\mathbf{U} = \{U_i\}_{i=1}^n$, $\mathcal{R} = \sigma(\mathcal{P}, S)$; and the utility assignments and the distributional weights λ_i are the strategies; and $U_i^{\mathcal{R}}$ is given as the Nash equilibrium of the constructed game. Then a coalition will determine its best response as the solution to

$$\max_{U_i} \sum_{S \in \Xi_C} \phi_S \ln g(\mathcal{P}, S, U_i) \tag{6.12a}$$

subject to (6.10),

$$f_{\mathcal{P},C}(U^{\mathcal{P}}) \leq 0 \tag{6.12b}$$

In addition, recalling that the distributional weights are identified only up to a multiplicative constant, we impose a normalization constraint:

$$\sum_{i \in C} \lambda_i \leq 1 \quad \forall C \in \mathcal{P}. \tag{6.12c}$$

Thus we form the Lagrangean function

$$\mathcal{L} = \sum_{S \in \Xi_C} \phi_S \ln \left[\sum_{i \in C} \lambda_i \left(U_i^{\mathcal{P}} - U_i^{\sigma(\mathcal{P},S)} \right) \right] - \mu f_{\mathcal{P},C}(\mathbf{U}^{\mathcal{P}})$$

$$+ \sum_{S \in \Xi_C} \mu_S f_{\sigma(\mathcal{P},S),S}(\mathbf{U}^{\mathcal{P}}) + \nu \left(1 - \sum_{i \in C} \lambda_i \right). \tag{6.13a}$$

Among the necessary conditions for the maximum are

$$\frac{\partial \mathcal{L}}{\partial \lambda_i} = \sum_{S \in \Xi_i} \frac{\phi_S (U_i^{\mathcal{P}} - U_i^{\sigma(\mathcal{P},S)})}{g(\mathcal{P}, S, \mathbf{U}^{\mathcal{P}})} - \nu \leq 0 \tag{6.13b}$$

$$\frac{\partial \mathcal{L}}{\partial U_i^{\mathcal{P}}} = \sum_{S \in \Xi_i} \frac{\phi_S \lambda_i}{g(\mathcal{P}, S, \mathbf{U}^{\mathcal{P}})} - \mu \frac{\partial f_{\mathcal{P},C}}{\partial U_i^{\mathcal{P}}} + \sum_{S \in \Xi_i} \mu_C \frac{\partial f_{\sigma(\mathcal{P},S),S}}{\partial U_i^{\mathcal{P}}} \leq 0. \tag{6.13c}$$

Note that the summations are over $S \in \Xi_i$, since for $T \notin \Xi_i$, the corresponding terms would not be defined.

From (6.13c), assuming an interior solution in $U_i^\mathcal{P}$, (as in Chapter 5, Section 5.2, and with the same rationale), we have

$$\lambda_i \sum_{S \in \Xi_i} \frac{\phi_S}{g(\mathcal{P}, S, \mathbf{U}^\mathcal{P})} = \mu \frac{\partial f_{\mathcal{P},C}}{\partial U_i^\mathcal{P}} - \sum_{S \in \Xi_i} \mu_C \frac{\partial f_{\sigma(\mathcal{P},S),S}}{\partial U_i^\mathcal{P}} \qquad (6.14a)$$

$$\lambda_i = \frac{\mu \dfrac{\partial f_{\mathcal{P},C}}{\partial U_i^\mathcal{P}} - \displaystyle\sum_{S \in \Xi_i} \mu_C \dfrac{\partial f_{\sigma(\mathcal{P},S),S}}{\partial U_i^\mathcal{P}}}{\displaystyle\sum_{S \in \Xi_i} \dfrac{\phi_S}{g(\mathcal{P}, S, \mathbf{U}^\mathcal{P})}}. \qquad (6.14b)$$

In (6.14a), essentially, the left-hand side of the equation is the "marginal benefit" of an increase in the utility assignment to i, and the right-hand side of the equation is the "marginal cost." We see that the "marginal benefit" increases with the distributional weight for i and with the bargaining power of groups in which he may participate, but decreases as the satisfaction of each such group increases; while the "marginal cost" is increased by an increase in the resource cost of increasing i's utility, $\mu \frac{\partial f_{\mathcal{P},C}}{\partial U_i^\mathcal{P}}$, and decreased by an increase in the productivity of an increase in the utility of i in meeting the rationality constraint for potential deviations, $\sum_{S \in \Xi_i} \mu_C \frac{\partial f_{\sigma(\mathcal{P},S),S}}{\partial U_i^\mathcal{P}}$. But variations in λ_i equilibrate these benefits and costs, and (6.14b) shows the equilibrium value for λ_i.

The term $\sum_{S \in \Xi_i} \frac{\phi_S}{g(\mathcal{P},S,\mathbf{U}^\mathcal{P})}$ may be thought of as an index of the overall influence of i on the decisions of C, an influence that increases with the bargaining power of any deviations in which i may participate and decreases with the gain of that deviation from continuing in C.

Suppose that, for $C \in \mathcal{P}$, none of the rationality constraints of potential deviating coalitions is binding; i.e., coalition C realizes a surplus with respect to every deviation that might disrupt it. Then $\mu_S = 0 \ \forall S \in \Xi_C$, and

$$\lambda_i = \frac{\mu \dfrac{\partial f_{\mathcal{P},C}}{\partial U_i^\mathcal{P}}}{\displaystyle\sum_{S \in \Xi_i} \dfrac{\phi_S}{g(\mathcal{P}, S, \mathbf{U}^\mathcal{P})}}. \qquad (6.14c)$$

Thus, in this case the distributional weight for i is the marginal cost of resources necessary to increase the utility assignment of i, adjusted (positively) for the incremental benefit of relaxing the feasibility constraint μ and negatively for the influence of i on the distributive process.

We cannot exclude the possibility that $\lambda_i = 0$. Suppose that

$$\mu \frac{\partial f_{\mathcal{P},C}}{\partial U_i^{\mathcal{P}}} \leq \sum_{S \in \Xi_i} \mu_C \frac{\partial f_{\sigma(\mathcal{P},S),S}}{\partial U_i^{\mathcal{P}}}. \tag{6.15a}$$

Then $\lambda_i = 0$. But then, by (6.13a), using the complementary slackness conditions and continuing to assume an interior solution in $U_i^{\mathcal{P}}$,

$$\mu \frac{\partial f_{\mathcal{P},C}}{\partial U_i^{\mathcal{P}}} = \sum_{S \in \Xi_i} \mu_C \frac{\partial f_{\sigma(\mathcal{P},S),S}}{\partial U_i^{\mathcal{P}}}. \tag{6.15b}$$

That is, the utility assignment to i is constraint-determined. In such a case, bargaining power plays no role, and this will play some part in the discussion below.

For now assume $\lambda_i > 0$, $\forall i \in C$. Then from (6.13b), we have

$$U_i^{\mathcal{P}} \sum_{S \in \Xi_i} \frac{\phi_S}{g(\mathcal{P}, S, \mathbf{U}^{\mathcal{P}})} = \sum_{S \in \Xi_i} \frac{\phi_S U_i^{\sigma(\mathcal{P},S)}}{g(\mathcal{P}, S, \mathbf{U}^{\mathcal{P}})} + \nu, \tag{6.16a}$$

$$U_i^{\mathcal{P}} = \sum_{S \in \Xi_i} w_S U_i^{\sigma(\mathcal{P},S)} + \frac{\nu}{\displaystyle\sum_{S \in \Xi_i} \frac{\phi_S}{g(\mathcal{P}, S, U^{U^{\mathcal{P}}})}}. \tag{6.16b}$$

where

$$w_S = \frac{\dfrac{\phi_S}{g(\mathcal{P}, S, U^{\mathcal{P}})}}{\displaystyle\sum_{T \in \Xi_i} \frac{\phi_T}{g(\mathcal{P}, T, U^{\mathcal{P}})}} \tag{6.16c}$$

so that

$$\sum_{T \in \Xi_i} w_S = 1. \tag{6.16d}$$

Thus, $U_i^{\mathcal{P}}$ will be a weighted average of the payoffs to i in the various deviations in which i might participate, with weights reflecting the proportions of i's influence on C that are derived from those deviations, plus a share of the surplus generated by the coalition above those weighted average payments. Once again we may think of (6.16a) as an equality of the marginal costs and the marginal benefits, with the left-hand side representing the benefit (in increasing the value of coalition (C) of a small increase in the distributional weight λ_i and the right-hand side the marginal cost of the same variation in λ_i. Thus, $\dfrac{\nu}{\sum_{S \in \Xi_i} \frac{\phi_S}{g(\mathcal{P}, S, \mathbf{U}^{\mathcal{P}})}}$, the share of the surplus to Agent i, is just large enough to balance the "benefits and costs" of raising the distributional weight λ_i.

Suppose that only individuals have bargaining power, i.e., $\phi_S = 0, \forall S \subseteq N, S \notin \mathcal{P} \ni |S| > 1$. Then

$$\sum_{S \in \Xi_i} \frac{\phi_S}{g(\mathcal{P}, S, \mathbf{U}^{\mathcal{P}})} = \frac{\phi_{\{i\}}}{\lambda_i(U_i^{\mathcal{P}} - U_i^{\sigma(\mathcal{P},\{i\})})}. \qquad (6.17a)$$

Equation (6.14a) becomes

$$U_i^{\mathcal{P}} = U_i^{\sigma(\mathcal{P},\{i\})} + \frac{\nu}{\mu \dfrac{\partial f_{\mathcal{P},C}}{\partial U_i^{\mathcal{P}}} - \sum_{S \in \Xi_i} \mu_C \dfrac{\partial f_{\sigma(\mathcal{P},\{i\}),\{i\}}}{\partial U_i^{\mathcal{P}}}} \qquad (6.17b)$$

with (6.14b), this yields

$$\lambda_i = \frac{\phi_{\{i\}}}{v}. \qquad (6.17b)$$

Moreover,

$$\sum_{j \in C} \lambda_j = 1 = \frac{\sum_{j \in C} \phi_{\{j\}}}{\nu}. \qquad (6.17c)$$

Thus

$$\nu = \sum_{j \in C} \phi_{\{j\}} \qquad (6.17d)$$

so that

$$\lambda_i = \frac{\phi_{\{i\}}}{\sum_{j \in C} \phi_{\{j\}}}. \qquad (6.17e)$$

In this special case, the distributional weight for i is simply the proportion of i's bargaining power in total bargaining power.

We see in this section, as in the previous chapter, that a value solution for NTU games corresponds to a determinate set of distributive weights for each embedded coalition, and so to a determinate value for each coalition. Thus, in practical applications, we might rely on the nucleolus or N-S nucleolus solutions of cooperative interactions in terms of TU games, understanding that this analysis presupposes an underlying NTU game with distributional weights determined along with the values. For applications in economics, however, a certain caution will be required. The values implied by a nucleolus or N-S-nucleolus solution need have little to do with market values or "value added" computed at market prices. It will, in addition or instead, reflect the bargaining influence of the members of the coalition, whether this influence arises from innate bargaining power or from strategic advantages in the form of alternative coalitions the individual might enter into.

6.6. Reflections

A shortcoming of this model, for direct application, is that the number of parameters is in principle quite large. Even for a four-person game there are 15 bargaining power parameters. It will not be possible to calibrate so many parameters, for example, for a macroeconomic model. But, in the phrase often wrongly attributed to Einstein, "Everything should be made as simple as possible but no simpler." All models choose to ignore some parameters and focus on others. The abstract consideration of this model may provide some important guidelines for the selection of simplifying assumptions that are not misleading. (1) Indirect bargaining power, operating through potential deviations that are never realized, cannot be ignored in applications of bargaining theory. (2) A group may be able to exercise bargaining power through threats that are not available to its members individually, so that, where group phenomena (such as an overall average wage level) are concerned, the group bargaining power should not be ignored, while individual bargaining power may be. (3) In such a case, the threat outcome or alternative cost of an individual deviating from the coalition may be irrelevant, within some range. (4) On the other hand, if the opportunity cost of the individual (individual rationality constraint) is above that range, it may determine the bargaining solution, as shown in the third example above.

6.7. Summary

This chapter has proposed a theory of imputation of values in the core of a characteristic function game, the N-S nucleolus, that, like that of Schmeidler (and Harsanyi) allows all possible deviations from the core to influence the imputation, and like Nash' assumes a smooth trade-off between the gains for different deviations, and further, like the Roth–Svejnar extension of Zeuthen–Nash bargaining theory, allow for exogenous differences in bargaining power. The result is a rather complex model, but one that can be solved by familiar methods of numerical optimization. Some examples of solutions of slightly different models illustrate the pitfalls of reducing multilateral bargains to bilateral bargaining theory, even when differences of bargaining power are abstracted from. The chapter has explored the application of the N-S nucleolus to the Morton–Pissarides–Diamond model of the determination of employment. This model, too, supposes that noncooperative strategies determine a cooperative match that implies an imputation problem, and it, too, has attempted to solve that problem by reducing the multilateral bargain among the members of the coalition called a firm to a series of bilateral bargains. It is suggested that a more satisfactory model might be obtained if the bargain were modeled explicitly as a multilateral bargain in which the work group as a whole have significant bargaining power independently of the bargaining power that workers may have as individuals. The chapter also extends this bargaining power solution to NTU games.

Chapter 7

Bargaining Power Biform Games

Brandenburger and Stuart (1996, 2007) have proposed "Biform Games" as a model suitable to many business applications. A Biform Game is a two-stage game, in which the first stage is noncooperative and the second is cooperative. The first stage is a noncooperative choice of strategies, but in place of a vector of payoffs the noncooperative game leads (depending on the strategies chosen) to a cooperative game played at the second stage. "However, the consequences of [the noncooperative] moves are not payoffs (at least not directly). Instead, each profile of strategic choices at the first stage leads to a second-stage, cooperative game." The cooperative game is represented in coalition function form (Brandenburger and Stuart, 2007, p. 538). The second-stage cooperative analysis relies on the theory of the core to set limits on the payoffs of the collaborating individuals. The game as a whole is solved by backward induction, with value solutions to the cooperative stage forming the payoffs in a reduced noncooperative game based on the first stage.

Since the core is based on a criterion of stability against deviations to new coalitions — i.e., stability in the face of competition — it seems an appropriate criterion for the cooperative stage of the game. Moreover, as Brandenburger and Stuart observe (2007, p. 547), the Biform Game approach avoids some of the shortcomings of more traditional cooperative game analyses: it does not presume efficiency of the outcome and allows externalities to be introduced in a natural way. On the other hand, however, as they note, (2007, p. 542) "there may be a range of values in the core." Thus, the payoffs in the first-stage noncooperative game would not be determinate. The solution to this problem may be influenced by bargaining power. Brandenburger and Stuart resolve this by means of an "index of optimism," α, whereby the decision-maker expects his payoff to be a weighted average of the lower and upper limits of the payoffs to him that are consistent with the core criterion. That is, letting v_1 be the upper limit and v_2

129

the lower, the decision-maker expects a payoff of $\alpha v_1 + (1 - \alpha)v_2$. But, as they observe, (p. 548) these indices may not be consistent. Chatain and Zemsky (2007) apply the Biform Game approach to the horizontal scope of the firm, imposing the assumption that (for two-person bargaining) the optimism indices of the bargainers sum to one, for institutional reasons.

Section 7.1 suggests a slight correction of the noncooperative stage of a Biform Game. Sections 7.3–7.6 explore some applications, and, with Sec. 7.2, will introduce the Bargaining Power Biform Game model with the N-S nucleolus solution at the second stage.

7.1. A Brief Digression on the Noncooperative Stage of a Biform Game

The literature on Biform Games recognizes that the core may not be unique in the cooperative stage of the game. A converse problem arises from the fact that the core of a cooperative game may be a null set (see Brandenburger and Stuart, p. 25). In that case, the payoffs in the reduced noncooperative game will not be defined.[1] Essentially, a solution in a Biform Game is a situation that is stable in two senses: in a cooperative sense (as the core is a concept of stability against the defection of some members of a coalition) and in a noncooperative sense (as Nash equilibrium can be thought of as stability against individual shifts of strategies). Accordingly, it will be necessary somewhat to modify the definition of the noncooperative game in normal form and the Nash equilibrium. The usual formal definition of a game in normal form is along these lines: the game comprises

(I) An index set of players, $N = \{1, 2, \ldots, n\}$.

(II) A set of strategy sets $\Sigma = \{\{\sigma_j\}_{j=1}^{k_1}, \{\sigma_j\}_{j=1}^{k_2}, \ldots, \{\sigma_j\}_{j=1}^{k_n}\}$.

(III) Letting $\mathbf{s}$ be any vector such that $s_i \in \sigma_i$, a payoff function $f : \mathbf{s} \to \Re^N$.

Then a Nash equilibrium is a vector $\mathbf{s}$ such that

(IV) $\forall i \in N$, $f(\mathbf{s}) \geq f(\mathbf{s}')$ where $s_j' = s_j \, \forall j \neq i, j \in N$.

---

[1] A qualification is that we might adopt a value solution such as the Shapley value or the nucleolus, which assigns values even in games for which the core is null. However, this seems inconsistent with the focus of the Biform Game literature on coalitions that do generate some "value added." Presumably coalitions that generate negative value added (which is the case if the core is null) will not be observed. However, that assumption needs to be part of the Biform Game formalism.

For present purposes, (ii) will become

(V) $f : \mathbf{s} \to (\Re^N \cup \{u\})$

where u is the state in which the payoffs are undefined; and in place of IV, we will characterize the stable strategies as a Cautious Nash Equilibrium (CNE), i.e.,

(VI) $\forall i \in N$, if $f(\mathbf{s}) \neq u$ and $f(\mathbf{s}') \neq u$, then $f(\mathbf{s}) \geq f(\mathbf{s}')$ where $s'_j = s_j \forall j \neq i, \ j \in N$.

For the purposes of this chapter and the next, condition (VI) characterizes the solution of the first-stage reduced game.

7.2. Bargaining Power Biform Games

Formally a Biform Game comprises

(I) An index set of players, $N = \{1, 2, \ldots, n\}$.

(II) A set of strategy sets $\Sigma = \{\{\sigma_j\}_{j=1}^{k_1}, \{\sigma_j\}_{j=1}^{k_2}, \ldots, \{\sigma_j\}_{j=1}^{k_n}\}$.

(III) A set of cooperative games, that is, value functions,

$$V = \{v_q : C \to \Re | C \subseteq N\}.$$

(IV) A function f from σ to V, where σ is the set of strategy vectors

$$\sigma = \{\sigma_{j_1}, \sigma_{j_2}, \ldots, \sigma_{j_n}\}.$$

We will define a Bargaining Power Biform Game as comprising (I)–(IV) in together with

(V) A power function, that is, a function $\phi(C) = \{\phi : C \to \Re | C \subseteq N\}$.

The function ϕ expresses the bargaining power in the possession of each coalition in the event that it threatens to deviate from an existing or proposed coalition. With this additional information we may compute the N-S nucleolus as the value solution of the cooperative game played at the second stage of the Biform Game.

7.3. Biform Games and Externalities

It has been remarked that the Biform Game approach can incorporate externalities and resulting inefficiency. In fact, externalities provide a good illustration of the Biform approach, as we see in the following example, Game 7.1. There are three Agents, a, b, and c. Agent a is considering an entrepreneurial entry. At the first, noncooperative stage, he can commit himself to one of two technologies. For this game,

(A) At the outset, each agent has wealth of 10.
(B) Technology 1 permits the production of 20 in a two-person coalition of the entrepreneur with one of the other two agents, and nothing in a singleton coalition. However, this commitment reduces the wealth each agent by 4, from 10 to 6. In the case of Agents b and c, this is a negative externality.
(C) Technology 2 permits the production of 10 with no effect on the pre-existing wealth of any agent.
(D) If b or c, or both, join a production coalition with a, then b or c or both experience an effort cost of 1.[2]

Clearly, Technology 2 is the more efficient, in that it yields a net product of 10 as against 8 for Technology 1. At the second stage of the game, a commitment to Technology 1 yields cooperative Game 7.1a, and a commitment to Technology 2 yields Game 7.1b. The games are shown in coalition function form in Table 7.1.

Both of these games have unique core imputations. For Game 7.1, the core imputation[3] is 25, 6, 6 for a, b, and c, respectively. For Game 7.2, the

Table 7.1.   Cooperative games in a biform externality game, Game 7.1.

	Game 7.1a	Game 7.1b
$\{a, b, c\}$	36	38
$\{a, b\}$	31	29
$\{a, c\}$	31	29
$\{b, c\}$	12	20
$\{a\}$	0	10
$\{b\}$	6	10
$\{c\}$	6	10

---

[2]This implies that the resulting cooperative game will not be superadditive.

[3]In Game 1, since the payoffs to a, b, and c (x_a, x_b, x_c) must total no less than 37, the grand coalition is not viable. However, in a partition of a two-person and a singleton coalition, such as $\{\{a, b\}, \{c\}\}$, $x_a \geq 25$, $x_b \geq 6$, and $x_c > 6$ and $x_a + x_b = 31$. These imputations are stable. The reasoning is similar for Game 7.1b.

Table 7.2.   The noncooperative stage of Game 7.1.

a	Technology 1	25
	Technology 2	19

core imputation is 19, 10, 10. Accordingly, the noncooperative first stage is the one-person game shown as Table 7.2.

Clearly, the Nash equilibrium of this game is unique and is that Technology 1 is chosen, although Technology 2 is efficient in the sense that it yields the greater total benefit. However, the negative externality generated by Technology 1 shifts bargaining power from Agents b and c, by reducing their disagreement payoffs, to Agent a in a way that more than compensates Agent a for the inefficiency of Technology 1.

Since the externality game has a unique core imputation in each case, its solution will correspond to the unique core imputation. In this case, the bargaining power of each coalition is determined by its rationality constraints, that is, by its competitive alternatives.

It may be of interest to revisit the "Coase Theorem" in the light of this discussion.[4] We may interpret the pre-existing wealth of the three agents as the value of an unpolluted tract of land that each of them owns, adjoining one another. The first-stage decision presupposes that Agent a has the right to develop has own property as he chooses, even though it results in the pollution of all three plots. The alternative property system would be one in which a plot of land cannot be polluted without the owners' permission. For Coase' reasoning, with the first property regime, Agents b and c could bribe a to adopt Technology 2. If each were to offer Agent a 3.5 from their pre-existing wealth, so that the payoff vector from Technology 2 would be 26, 6.5, 6.5, then all would be better off than is the case if Technology 1 were adopted. But this requires that Agent a's choice of the technology is part of a binding agreement among the three; that is, it requires that the first stage, like the second, is cooperative; in other words, it requires a fully cooperative solution, not a Biform Game solution.[5] Since one of

---

[4]It has long been known that the Coase theorem is not valid in a cooperative game theory analysis; see Aivazian and Callen (1981). Our concerns here are somewhat different, as Biform Games combine cooperative and noncooperative aspects.

[5]We observe in passing that the shift of property rights regimes would change the bargaining powers of the agents and coalitions, even in a fully cooperative model, since it would modify their competitive alternatives and their threats. (Indeed the first of these

Table 7.3.   Game 7.1′.

Coalitions	Game 7.1′
$\{a, b, c\}$	36
$\{a, b\}$	20
$\{a, c\}$	20
$\{b, c\}$	20
$\{a\}$	10
$\{b\}$	10
$\{c\}$	10

the objectives of the Biform Game approach is to explain the existence of persistent, inefficient externalities, this is an advantage of the Biform Game approach. For the second property regime, by contrast, it would be necessary to form the grand coalition if Technology 1 were to be chosen. The second-stage cooperative Game 7.1a would then be as shown in Table 7.3. Note that Game 7.1b is not affected by the difference in property rights, since it is not necessary for a to get the agreement of both b and c to adopt that technology.

In Game 7.1′, the core is not unique, and in fact any assignment of benefits that grants 10 or more to each player is within the core. Thus, unlike Games 7.1a and 7.1b, the bargaining power will determine the payoffs in this case. If the three have equal bargaining power in this symmetrical game, and coalitions of two or more have no bargaining power, they will divide equally. This would imply that, at the first stage, Agent a would choose Technology 2. The polluting Technology 1 will be preferable to Agent a only if he has overwhelming bargaining power. In the case of equal division of the net benefits, however, the payoffs will be 12 each, so that c makes side payments of 7 to each of a and b, comprising 1 to compensate for effort costs, 4 to compensate for the external cost, and 2 to share the surplus generated by the use of Technology 1. Even if Agent c has overwhelming bargaining power, he must grant payoffs of at least 10 to each of the other agents to satisfy their rationality constraint. So the payoff to Agent c from adopting Technology 1 cannot be more than 16. This yields the noncooperative game in Table 7.4. Clearly the solution to the Biform Game for the second property regime is that Technology 2 will be adopted.

---

was central to the result of Aivazian and Callen.) Since Coase was not concerned with bargaining power, this is not a critique of his idea. However, it is important as an instance of a principle that will be important from this point on: it shows how institutional details may determine persistent bargaining power within coalitions.

Table 7.4.   The noncooperative
stage of the game in Table 7.1.

a	Technology 1	$x \leq 16$
	Technology 2	19

All in all, then, the issue between Coase and the Biform Game analysis is an empirical one: is there evidence of persistent, inefficient externalities, or indeed of any noncooperative activity at all, in the actual economy? Is there evidence that differences in property rights influence the choice of technologies some of which generate externalities?

7.4.   Some Examples from the Recent Literature

In this section we consider some recent literature on Biform Games with applications in business strategy.

7.4.1.   *Brandenburger and Stuart (2007)*

Brandenburger and Stuart (2007) illustrate the Biform Game by the branded-ingredient game. This is a game between a supplier of a semi-finished material and a manufacturer whose product includes that material. There are two manufacturers, of which $F1$ is the stronger, in that customers' willingness to pay for a final product incorporating their product is greater. The supplier chooses at the first stage between a "status quo" strategy and a costly strategy of creating a brand image improvement, along the lines of Intel's "Intel inside" campaign. The brand improvement strategy would improve the willingness to pay for firm $F2$, the second manufacturer, but not that of $F1$, the first manufacturer. The supplier cannot supply both because of a capacity limitation. If at the first stage we have the supplier choosing the status quo, then the resulting cooperative game, Game 7.2a, is the three-person game shown in Table 7.5. In the table, S denotes the supplier, $F1$ the first manufacturer, and $F2$ the second manufacturer. The third column is the power function, which will be used in the N-S solution. Since the supplier cannot supply both firms, the core for this game is consistent with the coalition structure $\{\{S, F1\}, \{F2\}\}$ (that is, the supplier will find it more profitable to deal with $F1$.) Any imputation that awards zero to $F2$ and at least 2 to the supplier will be in the core for this game.

Suppose instead that at the first stage the supplier chose the Brand Improvement strategy. Then we have the three-person game shown in

Table 7.5. Game 7.2a. The branded ingredient game with a status quo strategy.

Coalitions	v	ϕ
$\{S, F1, F2\}$	8	0
$\{S, F1\}$	8	0
$\{S, F2\}$	2	0
$\{F1, F2\}$	0	0
$\{S\}$	0	1
$\{F1\}$	0	1
$\{F2\}$	0	1

Table 7.6. Game 7.6b. The branded ingredient game with a brand improvement strategy.

Coalitions	v	ϕ
$\{S, F1, F2\}$	8	0
$\{S, F1\}$	6	0
$\{S, F2\}$	2	0
$\{F1, F2\}$	0	0
$\{S\}$	0	1
$\{F1\}$	0	1
$\{F2\}$	0	1

Table 7.6, Game 7.2b. The payoffs in this case reflect (1) the cost of the brand improvement research and development to firm S and (2) the increase in the willingness-to-pay of the customers of firm $F2$. In this game, again, the coalition $\{S, F1\}$ will form, but core imputations are those that award 0 to $F2$ and at least 5 to S.

Once again, the first stage of this game is a one-person game. We see that the cores are indeterminate for both of these games. Thus, as Brandenburg and Stuart observe, a bargaining problem arises. Brandenburger and Stuart model bargaining by assuming that each player has confidence in his ability to bargain successfully that varies from 0 to 1. Denote this index of optimism by α. The supplier's expected payoff then is $\alpha V_1 + (1-\alpha)V_2$, where V_1 is the upper and V_2 the lower limit of his payoff in the core. Brandenburger and Stuart calculate that the supplier will choose the branded ingredient strategy unless $\alpha \geq \frac{3}{4}$. Suppose, for example, that S has $\alpha = \frac{1}{2}$. Then S expects that for Game 7.2a, he will get $(0.5)8 + (0.5)2 = 5$, while for

Game 7.2b, he will get $(0.5)8 + (0.5)5 = 6.5$. Note, however, that there is no requirement that all players have the same indices of optimism. Suppose that F1 has $\alpha = 0.8$. Then he expects a payoff in Game 7.2b. of $(0.8)3 + (0.2)0 = 2.4$. However, $6.5 + 2.4 = 8.9 > 8$, so the expectations of both agents cannot be met. Presumably there will be an impasse in the bargaining between S and $F1$ in this case, with some modification in their α values before a mutually agreed imputation is arrived at. The Brandenburger and Stuart "bargaining model" does not offer us any guidance on this. Thus, because of the possible inconsistency of the α values, the Brandenburger and Stuart model does not supply us with an unambiguous value solution for the cooperative stage of the game. It predicts the strategic solution in this case without ambiguity *because* the first-stage game is a one-person game, so that inconsistent beliefs about payoffs at the second stage do not affect the solution.

Now we may reconsider the Brandenburger and Stuart solution on the supposition that the second stage game is a BP game. As we have noted, a large number of parameters, one value of ϕ for each coalition, will have to be specified. In this case we specify them for comparability to the Brandenburger and Stuart discussion. Thus, for this example assume all assume that all agents have equal bargaining power as singleton coalitions, measured without loss of generality as 1, and all other coalitions have zero bargaining power. This is shown in Tables 7.5, 7.6 by the column headed ϕ. The solution to the status quo game, Game 7.2a. is shown in Table 7.7. For the brand improvement game, solution is in Table 7.8. The group rationality constraint for $\{S, F2\}$ is binding with a Lagrange multiplier of $\frac{1}{3}$, and this yields a much better payoff for the supplier. Thus, given equal bargaining

Table 7.7. The solution to Game 7.2a.

x_S	4
x_{F1}	4
x_{F2}	0

Table 7.8. The solution to Game 7.2b.

x_S	6
x_{F1}	2
x_{F2}	0

powers and supposing that the cost of brand improvement is 1, brand improvement is the supplier's first stage best response, even though the supplier does not deal with the firm that can benefit from it. This is consistent with Brandenburger and Stuart's discussion except for the details of the treatment of bargaining power.

Suppose, however, that the bargaining power of the supplier is less. If ϕ_S is $\frac{1}{3}$ or less, while $\phi_{F1} = \phi_{F2} = 1$, then the supplier gets the minimum payoff of 2 if he chooses the status quo. Suppose, yet again, that the supplier's bargaining power $\phi_S > 1.667$, with, again, $\phi_{F1} = \phi_{F2} = 1$, then with the brand improvement strategy $x_S > 5$. For example, with $\phi_S = 2$, $x_S = 5.333$. Then the status quo will be the supplier's best response. The greater bargaining power leads the supplier to choose the strategy that leads to the larger surplus, that is, the status quo strategy, since with greater bargaining power, the supplier gets a larger proportion of the larger surplus. Brandenburger and Stuart draw a similar conclusion (in their model the critical point comes when the supplier has three times the bargaining power of the buyer), and again, this solution differs from that of Brandenburger and Stuart only in the treatment of bargaining power. While this model seems more satisfactory in that there is no inconsistency among the bargaining power measures, pragmatically important differences may arise in other examples, as the next subsection shows.

7.4.2.   *Chatain and Zemsky (2007)*

Another application is found in Chatain and Zemsky's game of determining the horizontal scope of a firm. The first, illustrative example is a four-person game among a buyer, two specialist suppliers of services to the buyer, and one generalist. The first-stage strategy choice is the decision of the suppliers whether to enter the market or not, given an entry cost. If the generalist does enter, the result is a four-person cooperative game. This four-person game, Game 7.3a, is shown as Table 7.9. Note that the bargaining powers assumed here correspond to the optimism indices assumed by Chatain and Zemsky. Here, a is the buyer, b and c are the specialists, and d the generalist supplier of business services. The core for this game is consistent with coalition structure $\{\{a, b, c\}, \{d\}\}$. Since consequently the payoff to d is zero, the core allocation awards a minimum payoff of 180 to a. The remainder, 20, may be allocated among b and c in any way.

Chatain and Zemsky adopt Brandenburger and Stuart's optimism index and assume that the index for each participant is 0.5. They derive a value

Table 7.9.   Chatain and Zemsky's
four-person game, Game 7.3a.

Coalitions	v	ϕ
$\{a, b, c, d\}$	200	0
$\{a, b, c\}$	200	0
$\{a, b, d\}$	190	0
$\{a, c, d\}$	190	0
$\{b, c, d\}$	0	0
$\{a, b\}$	160	0
$\{a, c\}$	160	0
$\{a, d\}$	180	0
$\{c, d\}$	0	0
$\{b, d\}$	0	0
$\{b, c\}$	0	0
$\{a\}$	0	0.5
$\{b\}$	0	0.5
$\{c\}$	0	0.5
$\{d\}$	0	0.5

Table 7.10.   Chatain and Zemsky's
three-person game, Game 7.3b.

Coalitions	v	ϕ
$\{a, b, c, d\}$	200	0
$\{a, b\}$	160	0
$\{a, c\}$	160	0
$\{b, c\}$	0	0
$\{a\}$	0	0.5
$\{b\}$	0	0.5
$\{c\}$	0	0.5

solution of 190 for a, 5 for b and c, and 0 for d. Chatain and Zemsky then assume that there is an entry cost of 3, with the entry decisions by b, c, and d made noncooperatively in the first stage of the game. Agent d, expecting a payoff of zero if he does enter, finds that it is his best response not to enter, while b and c both can cover their entry costs and so their best response is to enter. However, this means that the cooperative game played in the Nash equilibrium case is not Game 7.3a but the three-person game shown in Table 7.10, Game 7.3b. The absence of competition from Agent d changes the core of this game and its value solution greatly. The core now includes any imputation that gives a at least 120. Using their optimism–pessimism approach, Chatain and Zemsky obtain a value solution of 160, 20, 20.

Table 7.11.   The first stage of Game 7.3.

			d			
		Enter			Don't	
		c			c	
Payoff order b, c, d		Enter	Don't		Enter	Don't
b Enter		$2, 2, -3$	$2, 0, 12$		$17, 17, 0$	$-3, 0, 0$
Don't		$0, 2, 12$	$0, 0, 87$		$0, -3, 0$	$0, 0, 0$

Thus, the first-stage game among b, c, and d is the three-person game shown in Table 7.11. This game has an unique Nash equilibrium, where b and c enter and d does not. Suppose, however, that b is very pessimistic, with $\alpha = 0.1$. Then the payoff he expects if he enters is at most $(0.1)(20) - 3 = -1$ if d does not enter and $(0.1)(5) - 3 = -2.5$ if d enters. Thus, the first-stage game implied in that case is quite different and has a unique Nash equilibrium at which b does not enter and c and d do. This indeterminacy is a consequence of the possible inconsistency of indices of optimism in the optimism-index approach.

We recall that the solution of Chatain and Zemsky was 100, 5, 5, 0. By contrast, the N-S nucleolus solution is 180, 10, 10, 0. In the N-S nucleolus computations, the only active constraint is the rationality constraint for $\{a, d\}$; which expresses Agent a's competitive alternative of contracting with d rather than b and c.

It is worth mentioning that application of Schmeidler's nucleolus to Chatain and Zemsky's problem *does* generate their solution. For Schmeidler's solution concept, increased gains for one deviation are not substitutable for decreased gains for another, so there is no normal tendency toward equal division of the gain. On the contrary, minimizing the maximum excess means that the gain for $\{a, d\}$ must be 10, not zero. The gain for $\{b, c\}$ then also is 10, and any shift of the imputation that increases the gain of $\{b, c\}$ would make the excess for $\{b, c\}$ greater than that for $\{a, d\}$, violating the minimization of the maximum excess.

Again, recall that in Game 7.3b, Chatain and Zemsky find a value solution of 160, 20, 20. By contrast the N-S nucleolus solution is 120, 40, 40. In this case the active constraints are the rationality constraints for $\{a, b\}$

and $\{a,c\}$; that is, the solution balances the threat that c will drop out of the grand coalition against the threat that b will do so.

Chatain and Zemsky go on to extend and apply their model in several ways, and we will not follow them in any detail. One possibility mentioned in the extensions is that it may pay firm a to subsidize the entry of potential competitors. This can be illustrated by the example considered here — if a could make a noncooperative commitment *ex ante* to subsidize d by a side payment of 4, on the condition that d enter the game, then a would increase his net equilibrium payout from 160 to 186 (using Chatain and Zemsky's value solution) or from 120 to 176 (using the N-S nucleolus solutions).

We see that Chatain and Zemsky's conclusions are quite sensitive to the *treatment* of bargaining power. This may arise because of the extreme asymmetry of their game, which assigns a preponderance of competitive power to Agent a. A bargaining model such as the N-S nucleolus, which allows for one agent or coalition's gain to be substituted for the game of another, tends to moderate that predominance of power.

7.5. Biform Games and Job Matching

The idea of costly matching of jobs with employees, central to so much recent work in macroeconomics including that honored by the Nobel Memorial prize of 2010, is an idea that fits well in the Biform Game model.[6] Commitments to post job openings or to search for a job are made noncooperatively at the first stage of the Biform Game, but once the match is made, the employer and employee enter into a cooperative relationship and play a cooperative game to determine the sharing of the benefits of their cooperation. To illustrate the idea, consider the three-person game in extensive form shown in Fig. 7.1. Agent a is an entrepreneur, in the sense that he may or may not take an initiative to form a coalition with b or c or both. The gross value of a two-person coalition of $\{a,b\}$ or $\{a,c\}$ is 10. (This is gross of matching costs.) The gross value of $\{a,b,c\}$ is 15. The cost of search is one, for Agents b and c, while for Agent a the cost of posting one job opening is 1 and the cost of posting two job openings is 2. Both b and c have outside options worth 2. (This might be an effort cost, so that the "outside

---

[6]See Chapter 6, Section 6.4, and discussion and references there.

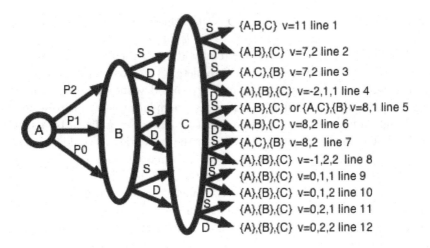

Fig. 7.1.   Game 7.4. A three-person matching game.

Table 7.12.   Coalition functions for some outcomes of the matching game.

Coalition	Line 1	Line 2	Line 5 (first interpretation)	Line 5 (second interpretation)
$\{a, b, c\}$	11	9	9	9
$\{a, b\}$	7	7	8	8 or 0
$\{a, c\}$	7	0	8	8 or 0
$\{b, c\}$	2	3	2	2
$\{a\}$	-2	-2	-1	-1
$\{b\}$	1	1	1	1
$\{c\}$	1	2	1	1

option" is avoidance of the effort.) If a coalition is formed then search costs are charged against the value of the coalition, and they are sunk costs. In the figure, the strategies are represented as $P0, P1, P2$ (post zero, one or two job openings) for Agent a and S or D (search or don't) for Agents b and c.

Table 7.12. shows some coalition value functions for various noncooperative strategy vectors. To illustrate how the results from the noncooperative stage of play are to be interpreted, consider first Line 1. The matching in the noncooperative step has brought all three players in the game together as a tentative coalition. However, they will proceed with the coalition only if they can agree on the payments. At the very least these must be in the core of the game, since otherwise some individual or group rationality constraints would be violated. If, for example, c were to reject the coalition and continue as a singleton, his payoff would be the outside option of 2 minus the search cost of 1 for a net of 1, while the coalition of $\{a, b\}$ could produce a value of 10 minus the search costs of 2 for a and 1 for b for a net of 7. Reasoning along these lines yields the coalition function in the second column of Table 7.12 for Line 1. For Line 2, c is not in touch with the other two and thus the value of $\{a, b, c\}$ is simply the sum of the values of $\{a, b\}$ and $\{c\}$ without any coordination of c's decisions with those of the other two. As before, $\{a, b\}$ can produce 10 but this is reduced by their search and posting cost of 3, for a net of 7. For $\{a, c\}$ there is no production and the value is the sum of c's outside option of 2 and a's posting cost of -2. For $\{b, c\}$ there is no production and the value of this pairing is the sum of their outside options less b's search cost, namely 3.

Line 5 is of particular interest, since it is the one outcome in which there are both employed and unemployed job searchers. It might be interpreted in either of two ways. First, it might be thought of as leading to a three-person coalition in which, however, only one of a and b will be employed and paid. For some reason, having committed himself by posting only one position, a cannot employ them both. (Perhaps he has solicited offers for only the product of one employee or has procured only complementary materials enough for one.) If, nevertheless, he can employ either of them at his choice, then we have the coalition function in the second column from the right. This game has a single allocation in the core — at which Agent a receives 7, b receives 1, and c receives 1 (regardless which of them is employed and which "unemployed"). This is an extreme case in which the bargaining power of employees depends on the existence or nonexistence of a reserve of unemployed, and in the case such a reserve exists, the employed do no better than the unemployed. But this is not in the spirit of the matching hypothesis, and indeed the evidence suggests that employees' bargaining power is not at all sensitive to the rate of unemployment, as noted in Chapter 6 in Section 6.4.2. The second interpretation is that, when a has posted one opening and both b and c search for a job, a is matched *at random* with *one or the other of them* so that Line 5 yields either $\{a, b\}, \{c\}$ or $\{a, c\}, \{b\}$ and a negotiates only with the one agent with whom he has been randomly matched. This sort of random matching plays a part in most if not all of the job matching models in the literature, and so is the interpretation adopted here. Of course, the difficulties may be with the small scale and stylized character of this example, but the second interpretation yields a surplus from the successful match, while the first does not, and this, too, is central both to the job matching literature and the Biform Game approach.

We may pause to characterize the cores of the 12 possible cooperative games that emerge from the noncooperative matching process in Fig. 7.1. In most cases this is trivial. For Line 1, where both agents b and c are matched to jobs, the core is characterized by $x_a \geq 3$, $x_b \geq 1$, $x_c \geq 1$. In all other cases (relying on the second interpretation of line 5) the core is determined by the constraint that no coalition distributes more than its value, together with the individual rationality constraints.

We will treat the second stage of the search game as a bargaining power game. The payoffs to the individual participants in the game will then depend on the bargaining power vector, ϕ, as well as the core constraints. The computed value solutions for several bargaining power vectors for the games at Lines 1, 2, 5, and 6 are shown in Table 7.13. In Table 7.13, for the second interpretation of Line 5, the payoffs for b and c are expected values assuming that the probabilities of an $\{a, b\}$ match and an $\{a, c\}$ match are equal. This is a simple average of Line 6 and Line 7. Lines 3 and 7 will reverse the order of the payments to b and c in Lines 2 and 6. At Lines 4 and 8–12 the payoffs are simply the values of the singleton coalitions.

In Table 7.13, for the second row (characterized by the bargaining power vector $0, 0, 0, 0, 1, 1, 1$) we impose the "common sense" assumption that only individuals have bargaining power and that the bargaining powers of the three individuals are equal. Rows 3-5 explore the possibility that the entrepreneur might have predominant or very predominant ($\phi \geq 10$) bargaining power. The last two rows explore the possibility that b and c together might have collective bargaining power. In the game partition implied by Lines 2–3 and 6–7 in Fig. 7.1 (e.g., Columns 3 and 6), the meaning of this assumption is a bit unclear, since no such coalition can form in those cases, and accordingly, no solution is shown in these cases.

Table 7.13.   **Bargaining power solutions of the matching game.**

ϕ	Line 1	Line 2	Line 5 first interpretation	Line 5 second interpretation	Line 6
$0, 0, 0, 0, 1, 1, 1$	$3, 4, 4$	$2, 5, 2$	$7, 1, 1$	$2.5, 3.75, 3.75$	$2.5, 5.5, 2$
$0, 0, 0, 0, 5, 1, 1$	$5.86, 2.57, 2.57$	$4.67, 2.33, 2$	$7, 1, 1$	$5.5, 2.75, 2.75$	$5.5, 2.5, 2$
$0, 0, 0, 0, 10, 1, 1$	$7.17, 1.92, 1.92$	$5.27, 1.72, 2$	$7, 1, 1$	$6.18, 1.9, 1.9$	$6.18, 1.81, 2$
$0, 0, 0, 0, 100, 1, 1$	$8.78, 1.11, 1.11$	$5.92, 1.08, 2$	$7, 1, 1$	$6.91, 1.55, 1.55$	$6.91, 1.09, 2$
$0, 0, 0, 5, 5, 1, 1$	$3, 4, 4$		$7, 1, 1$	$2.09, 3.95, 3.95$	
$0, 0, 0, 5, 1, 1, 1$	$3, 4, 4$		$7, 1, 1$	$0.71, 5.85, 5.85$	

Table 7.14. The reduced noncooperative matching game with equal individual bargaining power.

			Search		Don't	
			b		*b*	
Payoff order a, b, c		Search	Don't	Search	Don't
a	Post 2		$3, 4, 4$	$2, 2, 5$	$2, 5, 2$	$-2, 2, 2$
	Post 1		$2.5, 3.75, 3.75$	$2.5, 2, 5.5$	$2.5, 5.5, 2$	$-1, 2, 2$
	Post 0		$0, 1, 1$	$0, 2, 1$	$0, 1, 2$	$0, 2, 2$

Table 7.15. The reduced noncooperative matching game with overwhelming employer bargaining power.

			Search		Don't	
			b		*b*	
Payoff order a, b, c		Search	Don't	Search	Don't
a	Post 2		$8.78, 1.11, 1.11$	$5.92, 2.1.72$	$5.92, 1.72, 2$	$-2, 2, 2$
	Post 1		$6.91, 1.55, 1.55$	$6.91, 2, 1.09$	$6.91, 1.09, 2$	$-1, 2, 2$
	Post 0		$0, 1, 1$	$0, 2, 1$	$0, 1, 2$	$0, 2, 2$

Whatever the bargaining power vector, the value solution obtained enables us to construct a conventional noncooperative game representation of the first stage, by backward induction. For $\phi = 0, 0, 0, 0, 1, 1, 1$ and the second interpretation of Line 5, this is shown in strategic normal form as Table 7.14. The game in Table 7.14 has two Nash equilibria in pure strategies: at the upper left where the entrepreneur posts two openings and both worker agents search and a no-action equilibrium at the lower right at which no openings are posted and no job search takes place. In addition there is a mixed strategy solution in which *b* and *c* play "search" with probability 0.085 and a plays "post 1" with probability 0.23 and "Don't" otherwise. The probability that one position is posted and both *b* and *c* search is 0.0017. The probability that both search and a does not post a position is 0.0055, and the probability that a does not post and one of the agents searches is 0.12. Combining these, the expected value of the number of unemployed is 0.1327, that is, 6.6% of the potential labor force of 2.

In Table 7.15, we show the reduced noncooperative game corresponding to Line 5 of Table 7.13, that is, the assumption that the employer has overwhelming bargaining power. In this game "don't search" becomes a dominant strategy for b and c and the only Nash equilibrium is the no-action equilibrium. It appears that overwhelming bargaining power can operate to the disadvantage of the agent with the bargaining power, since it creates a situation in which other agents cannot recover their sunk costs of joining a coalition, so that opportunities to form a coalition cannot be realized.

7.6. A Corporate Governance Game

The examples considered so far are simplified in two ways. First, the first-stage noncooperative strategies are quite simple. However, noncooperative game theory allows for quite complex, contingent strategies, as, for example, schedules of payment that are contingent on the success of the joint operation. Second, (and particularly with respect to job matching) some of the participants in the noncooperative stage of the game may be pre-existing coalitions. We will illustrate these by Game 7.5, suggested by some of the recent literature on corporate governance (e.g., Tirole, 2001). We will first consider the game somewhat informally as a cooperative game. Risk neutrality is assumed, so that mathematically expected values are certainty equivalent. Two shareholders, a and b, each possess one of the nonhuman inputs necessary to undertake a project that could potentially generate a value of as much as 100. However, the value depends both on their joint strategy and on the state of nature. The coalition can choose actions X or Y and the state of nature may be Q or R. The payoffs are shown by Table 7.16. Agents a and b do not know the state of nature and they share a subjective probability estimate that the probability of each is $\frac{1}{2}$. However, they can include in their coalition a skilled manager c who is able to discover the state of nature.

If they do not recruit the expert to discover the state of the world, a and b will choose action Z, which gives them an expected value payoff of 70, rather than 60 from strategy Y. Thus, coalition $\{a, b\}$ can realize a value of 70. Now, suppose they recruit c, forming the grand coalition for this game. Then c can determine the state of nature and the grand

Table 7.16. Game 7.5. A joint game against nature.

	Q	R
Y	100	20
Z	60	80

coalition can choose a contingent strategy "If Q then Y; if R then Z."
This contingent strategy yields an *ex ante* gross expected value payoff of
90. This is a gross payoff because the discovery of the state of the world
has a subjective (effort) cost for c. Moreover, like the state of the world,
c's effort is not observable by a and b: only c knows whether he has gone
to the trouble or not. Thus, the agreement among the three will have to
be structured in such a way that c has some incentive to make the effort
to discover the state of the world. Whatever is left for a and b after the
payment to c is called "pledgeable income." Let us suppose (for Case 1)
that the effort cost is relatively small: it is 2. We also suppose that c has
an outside option so that $v\{c\} = 5$, which he must give up in order to join
the grand coalition,[7] while $v\{a\} = v\{b\} = 0$.

Suppose, nevertheless, that the contract awards c an uncontingent
payment of x_c. (Payouts to a and b will be denoted x_a and x_b.) Then,
if he does discover the state of the world, the net payoff to c will be $x_c - 2$,
whereas if he does not it will be x_c, so clearly it is not in his interest to
determine the state of the world (once his pay has been determined[8]). In
the literature on corporate governance, it is often proposed that he be paid
by a share of the profit,[9] but for this simple example, a share payment will
be inefficient.[10] Instead, let his contract grant a base payment of x_c plus a
bonus of ε if the total payoff is greater than 70 and zero otherwise. If he
does not discover the state of the world, his payoff will be

$$\frac{1}{2}x_c + \frac{1}{2}(x_c + \varepsilon) = x_c + \frac{\varepsilon}{2} \tag{7.1}$$

whereas if he does, it will be

$$x_c - 2 + \varepsilon. \tag{7.2}$$

---

[7] As a result of this, the game is not superadditive.

[8] While x_a and x_b are expected values of uncertain payouts we assume that the payoff
to c is fixed. This follows the corporate governance literature and is meant to support a
contrast between contingent and uncontingent payments to the expert.

[9] This idea has a long history! See, e.g., Mill (1879) Book 1, Chapter 9: "... it is not a
necessary consequence of joint stock management, that the persons employed, whether
in superior or in subordinate offices, should be paid wholly by fixed salaries. There
are modes of connecting more or less intimately the interest of the employees with the
pecuniary success of the concern. ... it is a common enough practice to connect their
pecuniary interest with the interest of their employers, by giving them part of their
remuneration in the form of a percentage on the profits."

[10] The minimal share payoff s that will assure c's effort in this case is one-tenth, and that
yields pledgeable income of 81. The reasoning is as follows. Let s be the share. If he does
not discover the state of the world his payoff is $70s$, while if he does it is $90s - 2$. Thus
to motivate c we must have $90s - 2 > 70s$, $20s > 2$, $s > \frac{1}{10}$.

**Table 7.17.   Game 7.5a
in coalitional form.**

Coalition	Value
$\{a, b, c\}$	88
$\{a, b\}$	70
$\{a, c\}$	0
$\{b, c\}$	0
$\{a\}$	0
$\{b\}$	0
$\{c\}$	5

It will be in c's interest to discover the state of the world if

$$x_c - 2 + \varepsilon > x_c + \frac{\varepsilon}{2} \tag{7.3}$$

i.e.,

$$\varepsilon > 4. \tag{7.4}$$

We suppose c is enlisted on the basis of that promise. The grand coalition will then realize a value of 88, net of the effort cost experienced by c. The coalition function for this game, Game 7.5a, is shown as Table 7.17. However, the solution will be constrained in that

$$x_c + \varepsilon \geq \varepsilon > 4. \tag{7.5}$$

That is,

$$x_a + x_b < 84 \tag{7.6}$$

and 84 is the "pledgeable income" of the coalition. But in this case constraint 7.5 will make no difference, since c's outside option and individual rationality require that

$$x_c + \varepsilon > 5. \tag{7.7}$$

Any imputation that gives a and b at least 70 and c at least 5 will be within the core for this game. It will be of interest to compute some common value solutions for this game. The Shapley value is $\{41, 41, 9\}$ while the nucleolus is $\{35.25, 35.25, 11.5\}$. We see that for the Shapley value $x_c + \varepsilon = 9 > 5$, for the nucleolus $x_c + \varepsilon = 11.5 > 5$, and in either case the division of c's payment between base salary and bonus can fall within a broad range

Table 7.18. Payoffs in Game 7.5a with the N-S solution and varying executive's bargaining power.

c's bargaining power	c's payoff	payoff to $\{a, b\}$
0.1	5.93	82.07
1	10.96	77.04
2	13.32	74.68
3	14.52	73.48
5	15.72	72.28
10	16.78	71.22
20	17.37	70.63
100	17.87	70.13

(for this very simple example). The Shapley value and the nucleolous do not admit any variation of bargaining power, however. Moreover, it is plausible that the manager's strategic situation might result in his having the greater bargaining power. Applying the N-S nucleolus, we may explore the consequences of a range of relative bargaining powers for Agent c, the manager. These are shown in Table 7.18, assuming that the bargaining power of $\{a\}, \{b\}$, and $\{a, b\}$ are one and that the bargaining powers of $\{a, c\}$ and $\{b, c\}$ are zero. Since the partition of interest is the grand coalition, its bargaining power will not affect the result. We note that whatever the bargaining power of c may be, a and b retain some surplus over their joint minimum payment, a "pledgeable income" of 70, and similarly, c gets some surplus over his minimum payment of 5, and moreover inequality 7.4 is satisfied, so that the only Nash equilibrium of the first-stage game is that c is recruited.

Considered as a cooperative game, however, this is not quite right. Agent c's failure to make the effort to discover the state of the world is a noncooperative decision. Considered as a Biform Game, we might treat the contingent payment schedule as the noncooperative strategy of the pre-existing coalition of a and b. Then a and b have a very large strategy set comprising all proportionate and bonus payment schedules. Conversely, then, c's noncooperative strategy is a contingent strategy of the form "If $x_c - 2 + \varepsilon > x_c + \frac{\varepsilon}{2}$, then make effort, otherwise do not." By making noncooperative commitments to these contingent strategies at the first stage, the agents impose constraints on the joint strategy of the coalition at the second stage (compare McCain, 1980) and so determine the cooperative game to be played. These contingent strategies are offers and demands made in the noncooperative hiring game. A large family of contingent

Table 7.19.   Game 7.5b in
coalitional form.

Coalition	Value
$\{a, b, c\}$	70
$\{a, b\}$	70
$\{a, c\}$	0
$\{b, c\}$	0
$\{a\}$	0
$\{b\}$	0
$\{c\}$	5

Table   7.20.   The   reduced   noncooperative
version of Game 7.5.

		c	
		Contingent effort	No effort
a and b	Satisfy	38.52, 38.53, 10.96	30.5, 30.5,4
	Don't	35, 35, 5	35, 35, 5

strategies will satisfy inequalities 7.3 and 7.7, so that Game 7.5 will bo the cooperative game played at the second stage. If not, the game will be Game 7.5b, as shown in Table 7.19. Here, as usual, we will need a value solution to define the first-stage noncooperative game. For simplicity, take just one representative strategy that satisfies the inequalities for a and b and one that does not, and just one contingent effort strategy for c that leads to effort if the inequalities are satisfied as against the strategy of making no effort regardless of payoff schedules; and adopt the N-S nucleolus with $\phi = 0, 1, 0, 0, 1, 1, 1$ as the payoffs. Then the reduced noncooperative game is as shown in Table 7.20. We see that the game has two Nash equilibria, at the upper left and lower right parts of the table of payoffs; but the lower right no-action equilibrium is weak and is Pareto-dominated by the Nash equilibrium at the upper left; moreover, the upper left equilibrium is a weakly dominant strategy solution. Thus, the upper left Nash equilibrium will be selected by a "trembling hand" refinement of Nash equilibrium. It also is the only strong or coalition-proof Nash equilibrium. (See, e.g., McCain, 2010, pp. 258–261, pp. 164–165.) If, instead, we adopt the Shapley value or the Schmeidler nucleolus as the value solution at the second stage, the details differ, but the same comments apply.

If the coalition is formed but c makes no effort, then Y and Z are chosen at random with probabilities $\frac{1}{2}$, so the expected value is 65, and that is the value of the coalition. Moreover, the chance of a realized value over 70 is $\frac{1}{2}$, so c's expected value payoff is $x_c + \frac{\varepsilon}{2}$. Taking these "counterfactuals" (in that no agent would enter into a bargain on these terms) the bargaining power solution would be 30.5, 30.5, 4.

In this example, the first-stage noncooperative game (which is the decision of a and b whether to recruit c) will always have a dominant strategy, that the expert is recruited, and the incentive constraint is satisfied, regardless of the treatment or distribution of bargaining power, since even in the last line of Table 7.16, with bargaining power of 100 attributed to c, the other two agents do better than they would in the alternative game in which c is not added to the coalition.

Let us instead suppose that the effort cost of discovering the state of the world is greater than 2. Suppose, for example, that it is 8. Then the value of the grand coalition is 82. The N-S nucleolus solution is $\{36.83, 36.83, 8.33\}$. We also have Shapley values of $\{39, 39, 4\}$ and the nucleolus value Schmeidler assignment is $\{36.75, 36.75, 8.5\}$. We note that the Shapley value for c does not satisfy the individual rationality constraint for c, since it is less than his outside option, but the nucleolus is individually rational in this version of the game.

In this version, however, constraint 7.4 becomes

$$\varepsilon > 16. \tag{7.4'}$$

Following the reasoning of the previous example, the pledgeable income of the coalition is 66, and in particular $x_a + x_b \leq 66$. We note that a and b can do better if they do not form the grand coalition with the expert and the state of the world is not discovered. Supposing that the coalition $\{a, b, c\}$ is formed, clearly $\{a, b\}$ can do no better than 66, which, if they have equal bargaining power, implies payoffs of 33, 33, 16. If instead we impose the constraint that $\varepsilon \geq 16$ in place of the rationality constraint for $\{a, b\}$ (supposing that, e.g., some regulation requires them to recruit c), then the solution is 31.125, 31.125, 25.75. Adopting the first of these as the upper limit of the payoffs for a and b, we obtain the first-stage game shown in Table 7.21. This game has two Nash equilibria, both in the bottom row, where a, b do not recruit c.

It appears that the need for an incentive compatible pay schedule for the expert manager conflicts with efficient organization in this case. However, there is another possibility. Suppose a and b approach c with the following proposition: first, c will make a payment into the coalition of 10. (In effect, x_c is -10.) Then c will be paid a bonus of 20.96 if the total payoff to the

Table 7.21.   The reduced noncooperative version
of Game 7.5 with greater cost of information.

		c	
		Contingent effort	No effort
a and b	satisfy	33, 33, 16	28.5, 28.5, 8
	don't	35, 35, 5	35, 35, 5

coalition is greater than 70 and nothing otherwise. If he does not discover
the state of the world the expected value of his payoff is

$$\frac{1}{2}(-10) + \frac{1}{2}(-10 + 20.96) = 0.48 \tag{7.8}$$

while if he does, it is

$$-10 + 20.96 = 10.96. \tag{7.9}$$

Since $10.96 > 0.48$, Agent c acting in his own interest will discover the state
of the world. The pledgeable income of the coalition in this case is 82, and
the expected value payoff to c is the N-S nucleolus value. The key to this
is that c is required to pay for the privilege of managing the coalition and
being paid a bonus if the results justify it. This again yields the first-stage
game in Table 7.20.

This is *not* equivalent to requiring that he be a shareholder. As a shareholder, his compensation would be a proportion of the realized coalition
value, and to motivate c to discover the state of the world, it would be
necessary to give him more than 40% of the shares, leaving a pledgeable
income of less than 54 for a and b.

The following lessons seem to follow from this example: first, that the
principle–agent problem is a mixture of cooperative and noncooperative
elements, although it is usually treated as a noncooperative interaction and
its cooperative aspects rendered as purely informal and intuitive motivation;
second, that the Biform Game approach can capture both aspects and
that our understanding of incentives for corporate executives and similar
principle–agent problems can be improved by doing so; and third, that
bargaining power, as in the different implicit assumptions about bargaining
power in the Shapley value and the nucleolus, can play a part in determining
the solution of the game.

7.7. Biform Games and Bargaining Power: Summary

Brandenburger and Stuart's Biform Games provide a coherent framework for integrating the cooperative and noncooperative aspects of many real-world interactions. Allowing for a first, noncooperative stage that determines what cooperative game will be played at the second stage, Biform Games can naturally incorporate such practically important phenomena as externalities and principle–agent interactions. However, Biform Games can often be solved only by reference to some theory of bargaining power. This chapter has applied the BP game approach at the second stage. In a case of a highly unsymmetrical game studied by Chatain and Zemsky, the choice of a bargaining theory can determine the outcome in ways that are important for application. Models with costly search and matching may also create bargaining problems, and can be represented naturally as Biform Games with the BP game as the cooperative second stage. The last chapter of the book will return to this topic.

Chapter 8

Intertemporal Cooperative Games: A Sketch of a Theory

Since the writings of Harrod (1939), Domar (1946), and Solow (1956), *intertemporal* modeling has become increasingly important in economics. Intertemporal models in economics capture two important points about real economic processes. First, some decisions yield results only over time. Thus, an investment in planting grapevines today will yield wine only in the future, and over many years in the future. Second, information may become available only over time. Thus, the weather that will influence the yield of the grapevines 50 years in the future will not be known for about 50 years.

Indeed, most modern macroeconomics is based on models of *dynamic stochastic general equilibrium*, which are intertemporal in both senses, and if many microeconomic models are atemporal, this is clearly understood as a counterfactual simplifying assumption. Intertemporal models in economics tend to retain the neoclassical assumption that human behavior is always noncooperative in large groups, and that large group processes are the decisive causes of economic phenomena. At the same time, game theory has become more and more influential. Within game theory, there are intertemporal noncooperative models, but consensus cooperative game theory remains resolutely atemporal. Coalition values are known for all time and coalitions form and distribute their surpluses once for all time. Nevertheless, in Erik Maskin's (2004) words, "we live our lives in coalitions." Coalitions, such as business firms, are key actors in almost all economic models. Thus, many noncooperative game theory models in economics are essentially noncooperative models of cooperative arrangements, and as such are logically inconsistent and incoherent.

It seems, then, that economics and game theory need a theory of intertemporal cooperative games. However, the barriers to such a theory are considerable. Intertemporal economic models usually rely on discounting of payoffs to present values and, more or less explicitly, on the Bellman (1957) principle. Long ago, however, Selten (1964) demonstrated that the Bellman Principle is not applicable to cooperative games in extensive form. Further, new information may from time to time alter the payoffs available to the agents. Implicitly, this will alter the values of the coalitions; however, for most cooperative-game solution concepts, these values are given data. Further, the players in the game will not be constant, as usually assumed; demographic development will remove some players while introducing new ones. Moreover, for pragmatic purposes, we need to allow for the fact that the same individual may be a member of different coalitions at different times, so that a simple partition of the agents into disjoint coalitions may not be a satisfactory representation of an intertemporal cooperative game.

The purpose of this chapter is, first, to explore some of the considerations that may arise in an intertemporal game, using informal and graphic examples, and second, to sketch a formal model of an intertemporal cooperative game and a concept of a presolution.

8.1.   A Primer for Cooperative Games in Extensive Form

Noncooperative models can incorporate the emergence of new information over time if the game is (at least partly) represented in extensive form. The once-for-all character of games in coalition function form is inherited from the normal form representation of the underlying games. Thus, to address intertemporal cooperative games, we must begin by discussing the identification of cooperative solutions in games represented in extensive form. There is, however, very little literature on this topic; indeed, so far as I have been able to discover, there is none since Selten's 1964 paper.

This section will discuss some elementary points about cooperative solutions in games in extensive form. These are not new. However, they are rarely made explicit, and it will be useful for some purposes in this chapter to make them so. Most cooperative game theory has followed the discussion of general games in von Neumann and Morgenstern (2004) in treating the underlying game only as a game in strategic normal form, but as we will see, there are some advantages in approaching cooperative games by discussing underlying games in extensive form.

Consider, for example, Game 8.1, shown in Fig. 8.1. This game has four possible outcomes, (2, 2), (5, 1), (1, 5), and (X, Y).

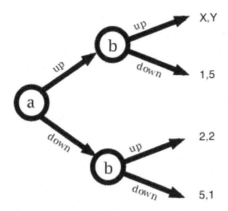

Fig. 8.1. Game 8.1.

(1) A cooperative solution is an agreement to choose coordinated strategies for mutual benefit. One possible condition we might set for such a
solution, and seemingly a minimal one, is that the outcome of the coordinated strategies should *not be Pareto-dominated* by any other feasible
outcome. Suppose, then, that $X = Y = 4$. (Then the payoffs in this
game correspond to the familiar social dilemma, although the information structure differs in that Agent b knows a's strategy.) In this case,
the outcome (2, 2) is excluded as it is Pareto-dominated by (X, Y).
However, none of the other three outcomes are Pareto-dominated; so
only (2, 2) can be eliminated at this stage.

(2) A further condition we might require for a cooperative solution is
symmetry. (Compare Chapter 2, Section 2.3.) For two-person games,
we might express symmetry in this way: suppose the game has the
property that whenever (Z, W) is a feasible outcome, then (W, Z) is
also a feasible outcome, and equal payoffs are feasible; then payoffs are
equal at a cooperative solution. Game 8.1 with $X = Y = 4$ has the
property of symmetry, so that $X = Y = 4$ would be the only eligible
cooperative solution for this game. Moreover, some further assumptions
can extend this limitation to games that it does not at first seem to
apply to. However, the symmetry assumption effectively excludes any
differences of bargaining power as between the individuals in the game.
To say that whenever (Z, W) is a feasible outcome, then (W, Z) is also
a feasible outcome is to say that for any threat Agent a may make,
Agent b has an equivalent threat to offset it. It is not clear, however,

that it is appropriate to assume away any differences of bargaining power between the parties to the negotiation.

(3) Yet another condition that we might consider imposing on a cooperative solution is *the Bellman principle*: that every decision is based only on the consideration of the alternatives available at the time the decision is made: that is, bygones are forever bygones. In a game such as Game 8.1, that means that the solution is subgame perfect, and with $X = Y = 4$, the solution at $(2,2)$ is the unique subgame perfect equilibrium. Thus, the Bellman principle leads to a solution that is Pareto-dominated. Since the exclusion of Pareto-dominated outcomes seems essential for cooperative solutions, the Bellman principle is inconsistent with a cooperative solution. (The influence of Selten's 1964 paper will be evident here.)

(4) Suppose that Agent a trustingly chooses strategy "up," an irreversible choice. Then Agent b opportunistically chooses "down." We would say that b has "defected" from the cooperative agreement. By doing so he has made himself better off. But this does not happen in a cooperative agreement, because such agreements are "enforceable." What does that mean? A game such as Game 8.1 might be nested (Tsebelis, 1990; McCain, 2010, Chapter 13) in a larger game such as Game 8.2, shown in Fig. 8.2.

The second move by Agent a is a threat move. By choosing "retaliate," Agent a reduces the benefit of b's defection from 5 to 3, which is sufficient to make it unprofitable for b to defect.

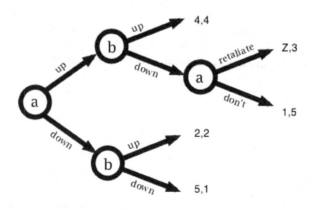

Fig. 8.2. Game 8.2.

But it may be unprofitable for a also. Suppose that $Z < 1$. Then (according to John Nash and on the basis of subgame-perfect equilibrium analysis) the threat will not be carried out, and Agent b will anticipate that it will not. On the other hand, if $Z \geq 1$, then the sequence "up," "up," would be consistent with the (noncooperative) subgame perfect equilibrium.

With $Z \geq 1$, we would have an instance of the "Nash program" in game theory. The program says that game theory should explicitly model the larger game in which the (cooperative) underlying game is nested, and in which compliance with the agreement becomes a noncooperative equilibrium. However, as against the Nash program, the following points may be made. (1) Experimental evidence indicates that people do have some capacity to commit themselves to carry out costly threats, especially where the commitment will, on the average, make them better off (as it does in this case). That is, the experimental evidence indicates that there is some probability that "retaliate" will be carried out even if $Z < 1$. (2) Human beings are at least as creative as they are rational and thus may find many ways to impose cheap retaliatory strategies on their counterparts. (3) It would be difficult if not impossible to model all of these strategies explicitly. Alternatively, for a cooperative analysis, we would simply assume that (in the absence of specific and justified assumptions to the contrary) cooperative agreements and threats will be carried out. (That is, in the language of the previous chapters, we adopt the assumption of costless cooperative rationality.)

Suppose instead that in Game 8.1, $X = 8, Y = 0$. In this case, none of the four outcomes are Pareto-dominated. We may suppose, however, that payoffs are in units of money and that money is divisible and transferrable. Then, again, we nest Game 8.1 in a bigger game, Game 8.3, as shown in Fig. 8.3. The second decision-node for Agent a is one at which a decides whether to make a "side payment" of z to b. Depending on the value of z, the payoffs to a and b can take infinitely many values, as shown in Fig. 8.4. Clearly, outcomes at $(1, 5)$, $(5, 1)$, and $(2, 2)$ are Pareto-dominated at this stage of the game. The strategies chosen by the two agents in a cooperative agreement will be those that maximize the total payoff (in this case, "up," "up.") This may not be trivial in a more complex game, but is something that is usually treated as given in cooperative game theory.

In Game 8.3, the payment of a side payment with positive z will never be subgame perfect. That is, an opportunistic Agent a would always choose "don't" at the last choice point. However, for cooperative games, as with Game 8.3, side payments may be essential to realize a mutual benefit from

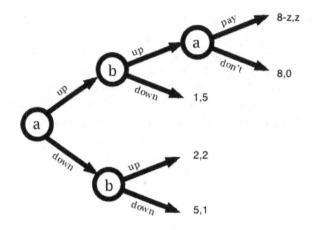

Fig. 8.3.   Game 8.3.

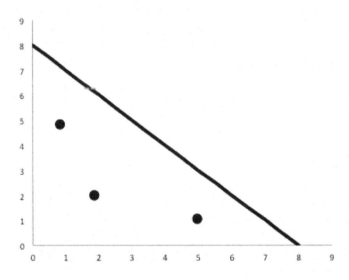

Fig. 8.4.   Payoffs in Game 8.3.

cooperation. Thus, side payments are strongly associated with cooperative games, and are never correctly associated with noncooperative games. Thus, moreover, Game 8.3 would be nested in a still bigger game with threat-enforcement strategies following the payment stage.

Nevertheless, the assumption we have made — that payoffs are in units of money — is problematic. It is subjective benefit, that is, utility, that actually motivates individuals in their choice of strategies. We might *assume* that transfers can be made in a way that is equivalent to transfers of utility. This assumption underlies the large literature on Transferable Utility (TU) games. An advantage of the TU assumption is that it makes the discussion of side payments relatively simple and natural.

Moreover, money may not be necessary for the transfer. Suppose that after the sequence "up, up" in Game 8.3, the agents enter into a wager that transfers the entire payment of 8 to Agent b with probability p. Then b's expected payoff is $8p = z$; that is, provided that b is risk-neutral, the probability $p = z/8$ is equivalent to a side payment of z. However, if b is risk-averse, then the wager is equivalent to something less than $8p$. In place of Fig. 8.4, we might have something like Fig. 8.5. In Fig. 8.5, curve qq' shows the upper limit of the utility of Agents a and b, expressed in money terms; that is, in the terms of welfare economics, it is a utility-possibility frontier.

It will be worthwhile to make the enforcement of the side payment in Game 8.3 explicit. Accordingly, consider Game 8.4 shown in Fig. 8.6. (In principle the subgame beginning from a's decision to pay or not would be appended also to all four of the outcomes of b's decisions between the "up" and "down" strategies, but these are left out for simplicity.) A cooperative solution for this game will correspond to an agreement on two specific *contingent* strategies; that is, on a specific strategy pair in the normal form game derived from Game 8.4. To attain a point on the utility frontiers in Figs. 8.4 and 8.5, a's contingent strategy will be of the form "Play 'up,' and

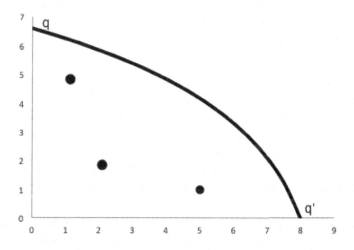

Fig. 8.5.   Certainty-equivalent payoffs in Game 8.3.

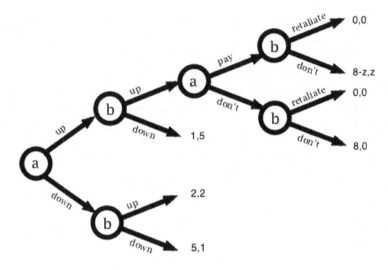

Fig. 8.6.   Game 8.4.

if b plays 'up' then pay z; else, don't pay." Correspondingly, b's cooperative strategy would be "If a plays 'up', then play 'up,' and if a plays 'don't' or pays less than z, then retaliate, else don't; else if a plays 'down,' then . . . ." In general, retaliation strategies will always be contingent.

This may suggest that the normal form expression is indeed the right basis for cooperative analysis, but, as we will see, the contrary is so, since it is precisely the construction of the contingent strategy from the game in extensive form — the step that the normal form elides — that is crucial for intertemporal cooperative analysis.

8.2.   Further Examples

Thus far, the examples given have not allowed for the emergence of new information nor for the economic value of time. This section will explore those things in a similar format of numerical and graphic examples. For simplicity value solutions in this section will be determined by Nash bargaining, so that with a given coalition value the surplus will be equally divided.

Consider, first, a game in which information becomes known during the play of the game, Game 8.5. The resolution of uncertainty is treated as a move by a fictitious player N (for Nature). In Game 8.5 shown in

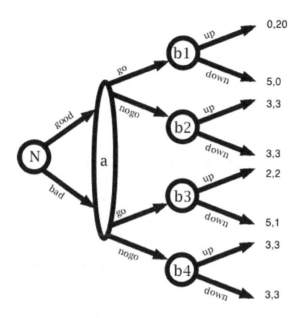

Fig. 8.7.   Game 8.5.

Fig. 8.7, Agent a must initiate some action in ignorance of the favorable or unfavorable conditions for it. As we see, Agent a chooses between the two strategies "go" and "no-go" at an information set. Once a makes the decision, the state of the world becomes known to both agents, so that b chooses the responding strategy "up" or "down" with knowledge of the state of the world, good or bad, as determined by nature at the outset. Then for a, with noncooperative play, the expected value of "go" is $2 - 2p$ (since it is subgame perfect for b to choose "up"), while for no-go a payoff of 3 is guaranteed. Thus the subgame perfect sequence is "no-go, either up or down indifferently."

Suppose, however, that a and b form a coalition. Assuming transferable utility, risk neutrality and equal bargaining powers, at $b1$ Agent b chooses "up" and at $b3$ "down," thus for the coalition the expected value of "go" is $6 + 14p$, while the expected value of "no-go" is, again, 3. Thus, " 'go,' and if good then 'up' else 'down' " is the cooperative strategy sequence. The coalition is formed before a's decision is made, and so in ignorance of the state of the world. If the probability p is known at that time, then b might offer a side payment to a in the amount of $12p - 2$, so that the expected values of payouts net of the side payment would be $3 + 7p$ each. Note that the side payment might be negative if p is quite small.

Suppose instead that the probability p is not known. Nevertheless, in this game, the noncooperative and cooperative contingent strategies are known, since the expected value of "go" is less than that of "no-go" in the noncooperative case, and greater in the cooperative case, regardless of p. (This would not be true in general.) Then the side payments might be determined by a contingent rule, such as "If the state of the world is good then b will pay 10 to a, else if the state of the world is bad a will pay 2 to b." Then the expected value of the payment from b to a is $-2 + 12p$. That is, an equal split of the total payout in each contingency is (of course) equivalent to an equal split in expected value terms.

Now, consider Game 8.6, shown in Fig. 8.8. In this game there are four possible states of nature, uw, uz, vw, and vz, in order of decreasingly favorable conditions for economic activity. At the outset, a, without any knowledge of the state of the world, must choose between two strategies, invest and don't. If "don't" is chosen, then b chooses "L" or "R" to determine payoffs in the same period and independently of the state of the world. If a chooses "invest," then there is no payoff until the second period. (The transition from the first to the second period is indicated in the figure by a vertical dashed gray line: everything to the right of this line takes place in the second period.) In the second period, the state of the world is known, and a and b play a social dilemma with payoffs that depend on the state of the world. It is assumed that the probabilities of the states of the world are public knowledge in both periods.[1] Further, it is assumed that payoffs at the second period are discounted to present value in period 1 by a factor δ. That is, Agents a and b both consider a unit of payoffs in period 2 equivalent to $\delta < 1$ of payoffs in period 1.

The noncooperative equilibrium for this game is "don't, R," regardless of the state of the world, since R is always a dominant strategy for b and, after eliminating the dominated strategy, "invest" can lead to no more than a payoff of 2 for a, while "don't" leads to a payoff of 3 for a.

The cooperative analysis will depend of the probabilities of the states of nature, but also on the time discount factor. Suppose that $\delta = 0.75$, the occurrence of w or z is independent on the realization of u or v, and $p(u) = 0.2$, $p(v) = 0.8$, and $p(w) = p(z) = 0.5$, so that $p(u, w) = p(u, z) = 0.1$

---

[1] For this example, it is possible that a's strategy choice might influence the probabilities of $\{uw, vw\}$ and $\{uz, vz\}$, as for example by choosing to plant a variety of seed that is better adapted to $\{uw, vw\}$ than to $\{uz, vz\}$, i.e., better adapted to conditions expected in the second period.

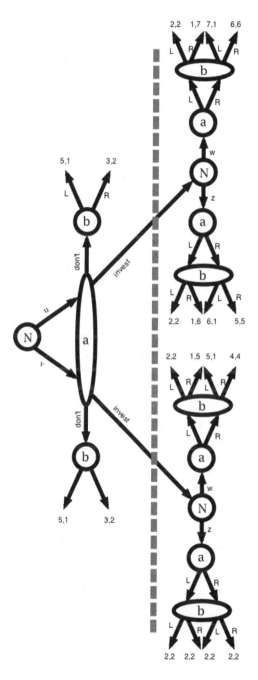

Fig. 8.8.   Game 8.6.

and $p(v, w) = p(v, z) = 0.4$. The coalition forms in the first period. Then, if a chooses "invest," the discounted expected value of the total payoffs to the coalition of both agents is no more than $E = 5.25$, while "don't" assures a coalition value of 6 in the first period. Thus, the cooperative strategy sequence in this case is "don't, L." But suppose instead that $\delta = 0.9$. Then $E = 6.3$, so that the cooperative strategy sequence is "invest, R, R." Assuming equal bargaining power, there will then be no need for side payments, since the payoffs are equal in each state of the world and, accordingly, in expected value terms as well.

Thus we see that — unsurprisingly! — the cooperative solution may depend on the discount factor. But we have assumed that both agents discount at the same rate. Suppose then that b discounts at the factor $\gamma = 0.75$ and a at $\delta = 0.9$. Thus the two disagree on the cooperative strategy. The implications of this depend on side payments, and specifically on the timing with which the side payments may be made.

First, suppose that a is able to make side payments at the first period, either from pre-existing wealth or by borrowing (although in the latter case it would be better to model the lender as part of the coalition). We might express this by saying that utility is transferable across time periods as well as between players. In particular, suppose a makes b an offer such as this: "a pays b X at period 1, and b pays a Y at period two, after the strategies 'invest, R, R' have been played."

Such a side payment offer can make it worthwhile for b to join the grand coalition even though the present value of the cooperative outcome in the next period has less value, at his own discount factor, than the noncooperative outcome. Since b values the payments at $X + 2.625 - \gamma Y$, the rationality constraint for b is

(1) $X + 2.625 - \gamma Y \geq 3$.

Equivalently

(2) $Y \leq \dfrac{X - 0.375}{0.75}$.

Similarly, the rationality constraint for a will be

(3) $-X + 3.15 + 0.9Y \geq 3$.

Equivalently

(4) $Y \geq \dfrac{X - 0.15}{0.9}$.

Indeed, since a strictly values second period payoffs more highly than b, it will be efficient for b to sell to a whatever his payoff in the second period may be. We then have side payments from b to a indexed by states as $Y_{uw} = 6, Y_{uz} = 5, Y_{vw} = 4, Y_{vz} = 2$. The discounted expected value of this contingent side payment for a is 3.15. Thus, before any payment from a to b, a's discounted expected value (and the value of the coalition) is 6.3, and a's expected discounted payout net of X is $6.3 - X$. Assuming equal bargaining power, then, $X = 3.15$. (This assumes, of course, that a begins the game with wealth sufficient to make such a payment at the first period.)

Notice that there is no intertemporal arbitrage here. Instead, b receives in period 1 the expected value that he gives up in period 2, *discounted at a's discount factor*. This is the case because the value of the coalition, and the surplus it generates, is determined by and so reflects a's discount factor. The extreme result in this example arises from the supposition that the rates of time preference are constant and payoffs are proportional to utility. If diminishing marginal utility of income in each period were introduced in some way, along with variable rates of time preference, then we might find an interior solution with equal rates of time preference at the margin. (The example does have an interior solution, but it proves to be a minimum, not a maximum.) It is an old idea in economics that differences in time preference, *per se*, are a basis for exchange, that is, for a cooperative coalition. Clearly, these considerations must be central to intertemporal cooperative games.

The two-person form of this example is a limitation, though, and explains why the solution does not have the form of arbitrage. In neoclassical economics, both a and b would have opportunities to form agreements with third parties, which would, in turn, also have many opportunities for agreements with others. Thus, a third party might offer to mediate the side payment between a and b and earn a profit by doing so. With many such third parties and many such deals, a zero-profit market interest rate would be formed so that there is no surplus at the margin. In such a case bargaining power would play no part. From the point of view of cooperative game theory, this third party is a member of the coalition and arbitrage in this sense is a property of the core of a game with large N. Conversely, if there are transaction costs of such intertemporal third party transactions, there will remain a surplus sufficient to offset the transaction costs, and so bargaining power may play a part, but within additional constraints determined by the market for arbitrage and the costs of transaction.

Second, suppose that players cannot make side payments before they have received payouts in the game. This means that, supposing that a chooses "invest," neither agent will be able to make side payments until the second period. Suppose then that b makes a side payment of Y to a at the second period. This will be discounted at a's higher discount factor. The discounted expected value to b of strategy "invest" is $2.625 - 0.75Y$ and to a it is $3.15 + 0.9Y$. Rationality constraints are

(5) $2.625 - 0.75Y \geq 2$
(6) $3.15 + 0.9Y \geq 3$

which yield

(7) $0.83 \geq Y \geq -0.167$

so that such a side payment seems feasible. We can derive a "utility-possibility frontier" that tells us the discounted present value payout to a as a function of that to b. It is

(8) $u_a = 6.3 - 1.1997u_b$.

In this case, the transferrable utility approach will not work, since the total value of the coalition, $u_a + u_b$, depends on the distribution. Applying Nash bargaining theory we derive a side payment of $Y = 0.33$, so that $u_b = 2.375, u_a = 3.45$, and the implied value of the coalition is 5.825.

Nevertheless, the example illustrates several key points. First, despite the inapplicability of the Bellman principle, payoffs to individuals can and must be discounted to present value. Side payments and strategic plans must be indexed by time period and the state of the world. If we assume that utility can be transferred between individuals and from one period to another, side payments will be made that maximize the discounted expected value of the coalition. Thus we can express the value of a coalition in discounted expected value terms. If, however, payments are not transferable over time and agents have differing time preferences, then TU methods cannot in general be used, so that an intertemporal game can only be an NTU game.

One complication not dealt with in any of these examples is demographic change, by which some players will be eliminated and new ones introduced depending on time period and the state of the world. Furthermore, for many practical purposes of economic theory, we would wish to allow for the possibility that an agent might participate in different coalitions in different time periods.

8.3. An Example with Demographic Change

Demographic change presents a complex challenge for cooperative game theory, since the players in the game may change from one time period to the next, sometimes depending on unpredictable natural events. Here is an example to illustrate some possibilities, Game 8.7. See Fig. 8.9. As before,

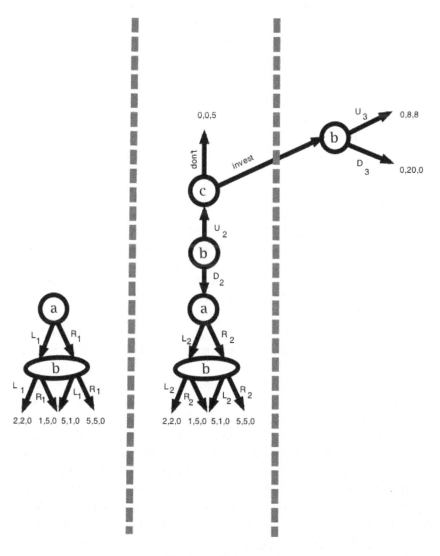

Fig. 8.9.   Game 8.7.

the transition from one time period to the next is indicated by a vertical dashed gray line.

For this example assume that $\delta = 1$ (or, alternatively, that all payoffs are discounted to present values at the first period). We see that Agents a and b are present in the first period, and are denoted as such by a_1 and b_1; a, b, and c are present at the second period and are denoted as a_2, b_2, and c_2; while only b makes a decision at period 3 and is denoted as b_3; but c_3 is present in the third period to receive side payments. For this example a coalition is a subset of $\{a_1, b_1, a_2, b_2, c_2, b_3 \, c_3\}$.

The noncooperative equilibrium for this game is, for a, "L_1, L_2"; for b, "L_1, D_2, L_2," and for c, "If b plays 'U_2' then play 'don't.' " For a cooperative solution, consider strategies:

(i) For a: R_1, R_2.
(ii) For b: (1) If a plays "R_1" then play "R_1," else play[2] "L_1," and in period 2 play "D_2" and if a plays "R_2" then play "R_2," else play "L_2."
(iii) For b: (2) If a plays "R_1" then play "R_1," else play "L_1," and in period 2 play "U_2" and if c plays "invest" then, in period 3, play "U_3."
(iv) For c: If b plays "U_2" in period 2 then play "invest."

Strategy (ii) yields payoffs $8, 8, 0$, while (iii) yields $4, 25, 0$. Neither Pareto-dominates the other. Assuming transferrable utility, however, (iii) dominates (ii) so that it would be the unique strategy array for the grand coalition, $\mathcal{G} = \{a_1, b_1, a_2, b_2, c_2, b_3, c_3\}$. Nevertheless, no side payments will be made to Agent a, since $\mathcal{G}$ with a positive side payment to a is dominated by $\mathcal{H} = \{b_2, c_2, b_3, c_3\}$. Note that strategies (i), (iii), (iv) are consistent also with the partition $\mathcal{P}_1 = \{\{a_1, b_1\}, \{a_2\}, \{b_2, c_2, b_3, c_3\}\}$ or with the partition $\mathcal{P}_2 = \{\{a_1, b_1, b_2, c_2, b_3, c_3\}, \{a_2\}\}$. For partition $\mathcal{P}_1$ we can compute a value solution using Nash bargaining theory, since it involves only two-person coalitions; and $\{a_1, b_1\}$ will pay $5, 5$, while $\{b_2, c_2, b_3, c_3\}$ will pay $10, 10$ (following a side payment in the third period of 10 from b to c) so that the overall payments are $5, 15, 10$.

Let us translate this into more common language. In the first period agents a and b collaborate for a joint profit of 6, which they share in equally. In the second period, however, a "shock" produces an opportunity for b to collaborate with c in a project that yields a profit of 10 (relative to b's outside option of 5 and c's outside option of 5) and, accordingly, b abandons his partnership with a to reaffiliate with c as a partner (partition $\mathcal{P}_1$) or lays

---

[2]Recall, in a cooperative solution, b can trust a's commitment to choose R_1.

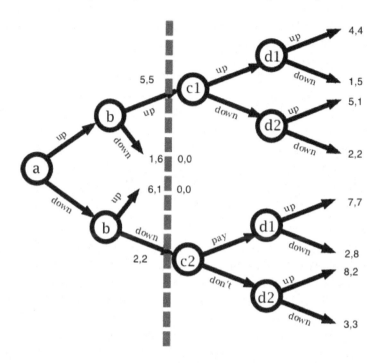

Fig. 8.10.   Game 8.8.

a off from "his" coalition and recruits *c* as a replacement (partition $\mathcal{P}_2$). Since *b*'s commitment to affiliation with *c* results in an outside option of 5 for *c*, the net surplus of their coalition is 10 and is divided equally.

One final example will be considered. Consider Game 8.8, Fig. 8.10. In this example, again $\delta = 1$. Payoffs in the first period are to a, b and in the second period to c, d. The key point is that the payoffs to *c* and *d* at the second period depend on the decisions of *a* and *b* at the first period.

In this game, in each period, a game is played that is a social dilemma except that there is no information set. Thus, the noncooperative play is "down, down, down, down" for payoffs of $2, 2, 3$, and 3. Cooperative play will depend on the coalitions that are formed. Consider the grand coalition $\mathcal{P}_1 = \{a, b, c, d\}$. Then we may have either "up, up, up, up" or "down, down, up, up." In either case the value of the grand coalition is 18. Now consider the partition $\mathcal{P}_2 = \{\{a, b\}, \{c, d\}\}$. In this case cooperative play sequences are "up, up"; "up, up" for values of 10 and 8. Now consider the partition $\mathcal{P}_3 = \{\{a\}, \{b\}, \{c, d\}\}$. Since there will not be cooperative play between *a* and *b*, the play sequence is "down"; "down"; "up, up," and the values of the coalitions are $2, 2, 14$. Assuming equal bargaining power, payoffs in $\mathcal{P}_2$ will be $5, 5, 4, 4$; that is, there will be no side payments. In the grand coalition any payment schedule that pays more than 8 to c and d will be dominated by the coalition $\{a, b\}$, so these payoffs will be made in the grand coalition as well.

This game illustrates two points. The first is that the game to be played in a later period may depend crucially on play in earlier periods. Each generation makes history only on the basis of the history made by earlier generations. Second, unlike the previous examples in this chapter, this game cannot directly be expressed as a coalition function. The value of $\{c, d\}$ depends on whether the coalition $\{a, b\}$ is formed. Thus, the coalition values might be expressed as a partition function. This issue is not particular to cooperative games in extensive form, and would arise in exactly the same way had the game been expressed in strategic normal form and the passage of time ignored. What this example illustrates more particularly, though, is that the partition function can be something of a historical document that captures interdependence over time, a key issue for intertemporal cooperative game theory.

This section has sketched a treatment of intertemporal games with a changing population, with two examples. The key suggestion is that these games can be expressed as games in coalition or partition function form. For this section, a coalition is a set of sets, each comprising an agent and a period in which the agent is active. Consequently, a coalition may be said to have a duration in time. Beginning from an expression of the games in extensive form, undominated strategy sets may be identified for each potential coalition or partition. Assuming that transfers can be made from one player to another that are equivalent to transfers of utility, and with appropriate discounting, the value of a coalition can be expressed in discounted expected value terms, discounted to values of the initial period.

For this approach, following von Neumann and Morgenstern, a strategy is a plan for each decision an agent may make at each decision node he may arrive at. For a cooperative approach with transferable utility, cooperative strategies will be chosen so as to maximize the value of each coalition in which the agent may at some time participate. Is it reasonable to suppose that agents may condition the strategies they choose in the current period on coalitions they might participate in in the future, with agents who do not yet exist? Surely the answer is a qualified yes. Rational agents surely will consider future opportunities to the best of their ability in making decisions in the present. Using the conventions of modern economic theory, so that decisions are indexed by the period and the state of the world, this consideration can be represented in such a way that coalitions have determinate values from the point of view of the present moment.

The next section will formalize these notions.

8.4. A More General Discussion

The proposition this chapter is intended to establish is an existential one: That there is a representation of intertemporal cooperative games with transferable utility that is consistent with the expression of the game in partition function form. As a matter of logic, an example is sufficient to establish existence, but in this case something more general is needed: to establish that such a representation can be made of some large class of demographic games along the lines of the examples in the previous section. For that purpose a constructive argument is required.

Games in extensive form are conventionally formalized as sequences of partitions of an abstract state-space, with a vector of payoffs corresponding to a state, following von Neumann and Morgenstern and Kuhn (1953). For present purposes a slightly more general (and unfortunately far less elegant) representation will be needed. The representation is needed to index decisions by time periods and to allow payoffs to occur in different periods, and sometimes before the conclusion of the game.

Accordingly, let Γ denote an intertemporal cooperative game in extensive form. The elements of Γ will include:

(1) A set of agents or players, $\mathbb{A} = \{\varnothing, 1, \ldots, i, \ldots\}$, which is countable and may or may not be finite.
(2) An ordered set of time periods $\mathbb{T} = \{1, 2, \ldots, t, \ldots\}$, which is countable and may or may not be finite.
(3) A set of decision points $\mathbb{D}$, with $d \in \mathbb{D}$, which is countable and may or may not be finite.
(4) A set of basic decision points $\mathbb{B} \subset \mathbb{D}$.
(5) A demographic correspondence $\mathbb{V} : \{(i, t) \ni i \in \mathbb{A}, t \in \mathbb{T}\}$.

Remark. If $(i, t) \in \mathbb{V}$, this means that Agent i is present at any decision point at time t; otherwise Agent i is not present to make decisions, receive payoffs or make or receive side payments.

(6) An agency function $\mathbb{A}(d) : \mathbb{D} \to \mathbb{V}$.

Remark. The agency function identifies d as taking place at a particular time, and as a decision made by a particular Agent a_i, with the qualification that if $a_i = \varnothing$, then d is a natural event.

(7) A successor relation $\mathbb{R}(d) : \mathbb{D}\backslash\mathbb{B} \to \{\mathbb{E} \subset \mathbb{D}|\mathbb{D} \notin \mathbb{E}\}$.

The following definitions and assumptions characterize the successor relation:

(i) Definition: If for $d \in \mathbb{D}$, there is no $d' \in \mathbb{D} \ni d \in \mathbb{R}(d)'$, then d is an *initiation point*.

(ii) Definition: Let $\rho \subset \mathbb{D} \ni \rho = \{d_1, d_2, \ldots, d_k\}$ and for $j = 2, \ldots, k, d_j \in \mathbb{R}(d_{j-1})$. Then ρ is an *action sequence*.

(iii) Assumption: Let ρ be an action sequence $\{f, d_2, \ldots, d_{k-1}g\}$, then *there is no action sequence* $\rho' = \{g, d_2', \ldots, d_{k'-1}f\}$.

Remark. That is, there are no decision cycles. In the terms of graph theory, treating decision points as vertices and pairs d_1 and $d_2 \in \mathbb{R}(d_1)$ as edges, a game is a tree.

(iv) Assumption: Let ρ be an action sequence $\{d_1, d_2, \ldots, d_k\}$. Let $t_1 \in \mathbb{A}(d_1), t_2 \in \mathbb{A}(d_k)$. Then $t_2 \geq t_1$.

Remark. That is, an action sequence cannot travel backward in time.

Trivial lemma: there is at least one initiation point. This will follow from (iii), (iv) and the indexation of time periods on the natural numbers. The earliest decision point (at least) must be an initiation point.

(v) $\forall d \in \mathbb{D}$, defining $\{d^*|\exists \rho$ an action sequence such that $d \in \rho$, $d^* \in \rho\}$, this set is finite.

Remark. That is, no action sequence continues forever without terminating at a basic decision.

(vi) Assumption: if ρ is an action sequence, $\rho = \{f, d_2, \ldots, d_{k-1}g\}$ and $\mathbb{A}(f) = (i, t_1)$, $\mathbb{A}(g) = (i, t_2)$, then $\forall t \ni t_1 \leq t \leq t_2$, $(i, t) \in \mathbb{V}$.

Remark. An agent's presence is continuous on action sequences.

(vii) Definition: $\mathbb{N} = \{d \in \mathbb{D}|\mathbb{A}(d) = (\varnothing, t)\}$

Remark. $\mathbb{N}$ is the set of natural events.

Further elements of an intertemporal game in extensive form are:

(8) A function from natural events to probability vectors, $\mathbb{P}(d) : \mathbb{N} \rightarrow \{\{\pi_1, \pi_2, \ldots, \pi_k\} | k$ is the cardinality of $\mathbb{R}(d).\}$

(9) A payoff function, $\mathbb{Z}(d)$. For $\mathbb{F} \ni \mathbb{B} \subset \mathbb{F} \subset \mathbb{D}$, $\mathbb{Z} : \mathbb{F} \rightarrow \{\{z_j\}_{j \in \mathbb{V}(d)} | z_j \in \Re\}$.

Remark. If $\mathbb{F} \backslash \mathbb{B} \neq \varnothing$, then this allows for payoffs at non-basic decision-points. In what follows $\mathbb{F}$ will denote the set of all decision points at which payoffs take place.

(10) A discount function, $\mathbb{G}$. $\mathbb{G} : \{\{i, t_1, t_2\} | i \in \mathbb{A}, t_1 \in \mathbb{T}, t_2 \in \mathbb{T}, t_1 < t_2\} \rightarrow (0, 1)$. Further, if $t_3 > t_2$, $\mathbb{G}(a_i, t_1, t_3) < \mathbb{G}(a_i, t_1, t_2)$.

Remark. Thus, if a payoff of z_i to i occurs at d with $\mathbb{A}(d) = (i, t_2)$ then the value a_i sets on the payment in period t_1 is $z_i \mathbb{G}(i, t_1, t_3)$.

To characterize noncooperative games and strategies, we require one additional element, an information correspondence.

(11) $\mathbb{H} : \mathbb{D} \rightarrow \{h \subset \mathbb{D}\} \ni$

    (a) $\forall h \in \mathbb{H}$, $d \in h(d)$.

    (b) $d^* \in h(d)$ and $\mathbb{A}(d) = (i, t) \Rightarrow \mathbb{A}(d^*) = (i, t)$.

Remark. An information set describes a particular agent at a particular time.

(c) $\forall h \in \mathbb{H}$, if $d^* \in h(d)$, $\mathbb{R}(d^*) = \mathbb{R}(d)$.

Remark. An agent knows what his options are, and cannot infer from that knowledge any restriction on the information set.

(d) $\forall h \in \mathbb{H}$, $\exists d^* \in \mathbb{D} \ni \forall d' \in h(d)$, $\exists \rho = \{d_1, d_2, \ldots, d_k\}$, ρ an action sequence, $\ni d^* = d_1$ and $d' = d_k$ and $\exists \rho' = \{b_1, b_2, \ldots, b_m\} \ni d^* = b_1$ and $d = b_m$.

Remark. That is, if d and d' are in the same information sets, there is a common initiation point from which both may follow by distinct action sequences. If $d^* = d$, then $h(d)$ has the single member $d^* = d' = d$, and we say that d is a perfect information decision point.

(e) If $d^* \in h(d)$ and $\rho = \{d_1, d_2, \ldots, d_k\}$, ρ an action sequence such that $d_k = d^*$, $\forall i \in \{1, 2, \ldots, k - 1\}$, $a_j \in \mathbb{A}(d_i) \neq a_i \in \mathbb{A}(d)$.

Remark. In other words, an agent will never forget his own past decision. This is Kuhn's assumption of perfect recall. (See, e.g., McCain, 2009, pp. 22–23.) Here we will follow Kuhn's assumption, or, in Kuhn's senses of the terms, we do not distinguish between players and agents.

Further definitions are

(viii) Let $\mathbb{H}_i = \{h(d) \in \mathbb{H} | \exists\, t \ni \mathbb{A}(d) = \{i, t\}\}$.

Remark. That is, $\mathbb{H}_i$ is the set of all information sets for Agent i. This will trivially include perfect information decisions.

(ix) Definition: A *pure strategy* for i is a function $\sigma \colon \mathbb{H}_i \to \mathbb{D} \ni \forall\, h(d) \in \mathbb{H}_i, \sigma(h) \in \mathbb{R}(d)$.

Remark. That is, a pure strategy is a function from information sets to successors.

(x) Definition: Let $\mathbb{S}_i$ be the set of all pure strategies for i and

$$\mathbb{S} = \{\{\sigma_1, \sigma_2, \ldots, \sigma_n\} | \sigma_i \in \mathbb{S}_i\}.$$

We may now characterize a noncooperative game in normal form quite conventionally as a function from sets of pure strategies to sets of expected value payoffs for the agents in the game.

In order to characterize cooperative decisions, we will require the following additional elements:

(12) A set of side payments, $\mathbb{P}(d)$. For $\forall\, d \in \mathbb{F}$,

$$\mathbb{P}(d) = \{\{p_{i,j}\}_{(i,t)\in \mathbb{V}(d)}\}_{(j,t)\in \mathbb{V}(d), j\neq i} \ni p_{i,j} \in \Re, \quad 0 \le p_{i,j} \le z_i.$$

Remark. For $a_i \in (i,t) \in \mathbb{V}(d)$, the net payoff is

$$y_i = z_i - \sum_{\substack{j\in \mathbb{V}(d) \\ j\neq i}} p_{i,j} + \sum_{\substack{j\in \mathbb{V}(d) \\ j\neq i}} p_{j,i}.$$

(13) Assumption: $\exists\, \delta \in (0,1) \ni \forall (a_i, t_1, t_2), \mathbb{G}(a_i, t_1, t_2) = \delta^{t_1 - t_2}$.

Remark. This is a simplifying assumption stating that there is an invariant common rate of time discount. This assumption, like transferable utility, is necessary in order to obtain an unambiguous value for a coalition.

(14) A coalition is a finite set $\mathbb{C} \subset \mathbb{V}$.

Remark. Thus, the membership of a coalition may differ from one time period to another. Moreover, a coalition will have a limited existence in time. This encompasses some probability that an individual may die at the end of a particular time period (as death is a natural event) as well as an example in which an agent leaves the coalition in order to accede to another or an example in which the coach suspends a player for one game because of the player's decision to skip a practice.

(15) For $\mathbb{C}$, let $T_1(\mathbb{C}) = \min_{d \ni \mathbb{A}(d) \in \mathbb{C}} t \in \mathbb{A}(d)$ and $T_2(\mathbb{C}) = \max_{d \ni \mathbb{A}(d) \in \mathbb{C}} t \in \mathbb{A}(d)$.

Remark. These are the earliest and latest dates of a coalition's existence as such.

(16) Given $\mathbb{C}$, $\mathbb{M}(\mathbb{C}) = \{i | \exists t \ni (i, t) \in \mathbb{C})$.

Remark. This is the membership of the coalition, encompassing all periods of its existence.

(17) Given $\mathbb{C}$, let $\mathbb{C}^\circ = \{d | \mathbb{A}(\mathbb{D}) \in \mathbb{C}, d \in (\mathbb{B} \cap \mathbb{N})\}$.

Remark. $\mathbb{C}^\circ$ is the set of decision points under the control of $\mathbb{C}$ which are not basic.

(18) (a) For $\mathbb{C}$, let $\mathbb{J}_\mathbb{C}(d)$ be a function $J_\mathbb{C}: \mathbb{C}^\circ \to \{d^* \in \mathbb{R}(d)\}$.

Remark. That is, $\mathbb{J}_\mathbb{C}$ maps decision points controlled by $\mathbb{C}$, that are neither basic nor natural events, into their successors.

(18) (b) For $\mathbb{C}$, let $\mathbb{K}_\mathbb{C}(d)$ be a function $\mathbb{K}_\mathbb{C}: \{\mathbb{C} \cap \mathbb{F}\} \to \{\mathbf{p} \in \mathbb{P}(d)\}$.

Remark. That is, $\mathbb{K}_\mathbb{C}$ maps decision points controlled by $\mathbb{C}$ that have payoffs into a vector of side payments at those decision points.

(19) For $\mathbb{C}$, a *joint strategy* $\mathbb{Y}_\mathbb{C}$ is a pair of functions $\mathbb{J}_\mathbb{C}$ and $\mathbb{K}_\mathbb{C}$.

(20) Now, given $\mathbb{C}, \mathbb{J}_\mathbb{C}$ and $\mathbb{K}_\mathbb{C}$, let ρ be an action sequence such that $d \in \rho \cap \mathbb{C}^\circ \Rightarrow \mathbb{J}_\mathbb{C}(d) \in \rho$. Then ρ is a coordinated action sequence for $\mathbb{C}$.

Remark. A coordinated action sequence is a conditional plan, where the conditioning events are natural events that precede decision points under the control of $\mathbb{C}$. Thus, given $\mathbb{J}_\mathbb{C}$, a probability can be attached to each coordinated action sequence by elementary compounding of probabilities. Without further formal discussion, denote

(21) The probability of ρ given $\mathbb{Y}_\mathbb{C}$ is π_ρ and $E(i, t, \mathbb{Y}_\mathbb{C})$ is the expected value of payoffs to i, net of side payments.

(22) Let $\mathcal{P}$ be a partition of $\mathbb{V}, \mathbb{C} \in \mathcal{P}$, and for $\mathbb{C}' \in (\mathcal{P} \backslash \{\mathbb{C}\})$, fix $\mathbb{Y}_{\mathbb{C}'}$ as the coordinated strategy of each coalition in $\mathcal{P}$ other than $\mathbb{C}$. For $\mathbb{C}$ consider joint strategy $\mathbb{Y}_\mathbb{C} = \{\mathbb{J}_\mathbb{C}, \mathbb{K}_\mathbb{C}\}$. Let $\boldsymbol{\rho} = \{\rho_m\}$ be an

enumeration of the coordinated action sequences derived from $\mathbb{Y}_{\mathbb{C}}$. Then let $E_{\mathbb{Y}_{\mathbb{C}}}(i) = \sum_{\substack{d \in \rho_m \cap \mathbb{K}_{\mathbb{C}} \\ \rho_m \in \rho}} \pi_\rho \delta(a_i, t_t, t_2) y_i$, where $t_1 = T_0$, $t_2 \in \mathbb{A}(d)$, $\delta \in \rho$, and y_i is evaluated as in expression 12.

(23) Let $E(\mathbb{Y}_{\mathbb{C}}) = \sum_{i \in M(\mathbb{C})} E_{\mathbb{Y}_{\mathbb{C}}}(i)$.

Lemma: $E(\mathbb{Y}_{\mathbb{C}}) = \sum_{i \in M(\mathbb{C})} E_{\mathbb{Y}_{\mathbb{C}}}(i) \sum_{\substack{d \in \rho_m \cap \mathbb{K}_{\mathbb{C}} \\ \rho_m \in \rho}} \delta^{t_2 - t_1} z_i$.

Proof.

$$E(\mathbb{Y}_{\mathbb{C}}) = \sum_{i \in M(\mathbb{C})} \sum_{\substack{d \in \rho_m \cap \mathbb{K}_{\mathbb{C}} \\ \rho_m \in \rho}} \pi_\rho \delta(a_i, t_t, t_2) y_i$$

$$= \sum_{i \in M(\mathbb{C})} \sum_{\substack{d \in \rho_m \cap \mathbb{K}_{\mathbb{C}} \\ \rho_m \in \rho}} \delta^{t_2 - t_1} \left[z_i - \sum_{\substack{j \in V(d) \\ j \neq i}} p_{i,j} + \sum_{\substack{j \in V(d) \\ j \neq i}} p_{j,i} \right]$$

$$= \sum_{i \in M(\mathbb{C})} \sum_{\substack{d \in \rho_m \cap \mathbb{K}_{\mathbb{C}} \\ \rho_m \in \rho}} \delta^{t_2 - t_1} z_i$$

$$+ \sum_{\substack{d \in \rho_m \cap \mathbb{K}_{\mathbb{C}} \\ \rho_m \in \rho}} \delta^{t_2 - t_1} \sum_{i \in M(\mathbb{C})} F_{\mathbb{Y}_{\mathbb{C}}}(i) \sum_{j \in V(d)} (p_{i,j} - p_{i,j})$$

Clearly, the second term is zero.   □

(24) Let $\mathcal{P}$ be a partition of $\mathbb{V}, \mathcal{P} = \{\mathbb{C}_1, \ldots, \mathbb{C}_M, \ldots, \}$ and let $\mathbb{Y} = \{\mathbb{Y}_1, \ldots, \mathbb{Y}_M, \ldots, \}$ be a set of joint strategies for the coalitions in $\mathcal{P}$. The strategies $\mathbb{Y}_M$ may be mixed strategies as well as pure strategies defined at 19. Then $\mathbb{Y}$ is *considerable* iff

$$\exists \{\mathbb{Y}_1, \ldots, \mathbb{Y}'_M, \ldots, \} = \mathbb{Y}' \neq \mathbb{Y}$$

$$\ni \exists i \in \mathbb{A} \ni \sum_{\mathbb{C}_M \ni i \in M(\mathbb{C}_M)} [E(i, 1, \mathbb{Y}'_M) - E(i, 1, \mathbb{Y}_M)] > 0$$

$$\Rightarrow \exists a_j \in \mathbb{A} \ni$$

(i) $\exists \mathbb{C}_K \in \mathcal{P} \ni i \in M(\mathbb{C}_K), j \in M(\mathbb{C}_K)$.

(ii) $\sum_{\mathbb{C}_M \ni j \in M(\mathbb{C}_M)} [E(j, 1, \mathbb{Y}'_M) - E(j, 1, \mathbb{Y}_M)] < 0$.

Remark. *Considerability* is a restricted Pareto optimality condition. Here is the reasoning in ordinary-language terms. Suppose that a coalition $\mathbb{C}_M$, by shifting its joint strategy from $\mathbb{Y}_M$ to $\mathbb{Y}'_M$, can make Agent i better off, and moreover it does so without any cost to any agent either in $\mathbb{C}_M$ or in any other coalition in which i may participate in the future. Then i will advocate the change, and has no reason to anticipate either resistance or retaliation either in $\mathbb{C}_M$ or in any future coalition in which he may participate. In such a case it is reasonable to suppose that the shift in joint strategies will occur; $\mathbb{Y}_M$ will not be considered as a candidate for the joint strategy of the coalition.

Consider, in particular, the special case in which $\mathbb{C}_K = \mathbb{C}_M$, i.e., $j \in \mathbb{C}_M$. Applying the definition to this special case we see that the condition of considerability subsumes the condition that the joint strategy is Pareto-optimal among the members of the coalition. Consider instead the special case with $\mathbb{C}_K \neq \mathbb{C}_M$ but $j = i$. We see that the condition of considerability subsumes the idea that Agent i will not support strategies that will result in his being disproportionately worse off in some future coalition. All in all, it may seem that the condition of considerability extends cooperative decisions beyond the scope of the coalition, and this is so just to the extent that the same agent may, at different times, participate in different coalitions. Nevertheless, a considerable array of joint strategies needs not be Pareto-optimal, and conversely. For consider a Pareto-optimal array of joint strategies $\{\mathbb{Y}^0{}_1, \dots, \mathbb{Y}^0{}_M, \dots, \} = \mathbb{Y}^0$. It follows that, for any $\mathbb{Y}'_M$, supposing $i \in \mathrm{M}(\mathbb{C}_M)$ and

$$\sum_{\mathbb{C}_M \ni i \in \mathrm{M}(\mathbb{C}_M)} [E(i, 1, \mathbb{Y}'_M) - E(i, 1, \mathbb{Y}_M)] > 0, \quad \exists\, j \in \mathrm{A} \ni$$

$$\sum_{\mathbb{C}_M \ni j \in \mathrm{M}(\mathbb{C}_M)} [E(j, 1, \mathbb{Y}'_M) - E(j, 1, \mathbb{Y}_M)] < 0,$$

if nevertheless, for all such j, $i \in \mathbb{C}_M$, $\mathbb{C}_K \neq \mathbb{C}_M \Rightarrow j \notin \mathbb{C}_M$, then condition (i) is not fulfilled, and so $\mathbb{Y}^0$ is not considerable. Conversely, suppose $\mathbb{Y}$ is considerable but $\mathbb{Y}^0{}_M \neq \mathbb{Y}_M$, $\mathbb{Y}^0{}_K \neq \mathbb{Y}_K$, and $\mathbb{Y}^0$ Pareto-dominates $\mathbb{Y}$. Then $\mathbb{Y}$ is clearly not Pareto-optimal. This can occur because the shift from $\mathbb{Y}$ to $\mathbb{Y}^0$ requires coordinated shifts of strategy by two (or more) coalitions that may or may not have any members in common.

(25) Theorem: Consider the noncooperative game in normal form comprising the coalitions of $\mathcal{P}$ as players, the joint strategies $\mathbb{Y}_M$ as strategies, and $E(\mathbb{Y}_C)$ as payoffs. Then $\{\mathbb{Y}_1, \dots, \mathbb{Y}_M, \dots, \}$ is considerable iff it is a Nash equilibrium of the noncooperative game.

Proof.

(i) If $\mathbb{Y}$ is considerable then it is a Nash equilibrium. Let $\{\mathbb{Y}_1, \ldots, \mathbb{Y}_M, \ldots, \}$ be considerable and contrast it with $\{\mathbb{Y}_1, \ldots, \mathbb{Y}'_M, \ldots, \}$, such that $\mathbb{Y}'_M \neq \mathbb{Y}'_M$. Then $E(\mathbb{Y}_M) \geq E(\mathbb{Y}'_M)$. For suppose otherwise: Then the payouts to all of the members of $\mathbb{C}_M$ can be increased without any change to the payments to players in any other coalition, in violation of (25). Thus, $\{\mathbb{Y}_1, \ldots, \mathbb{Y}_M, \ldots, \}$ is at least a weak Nash equilibrium.

(ii) If $\mathbb{Y}$ is a Nash equilibrium then it is considerable. Since $\mathbb{Y}$ is a Nash equilibrium, $\mathbb{Y}'_C \neq \mathbb{Y}_C \Rightarrow E(\mathbb{Y}'_C) \leq E(\mathbb{Y}_C)$. It then follows that

$$E_{\mathbb{Y}_C}(i) > E_{\mathbb{Y}'_C}(i) \Rightarrow \sum_{\substack{j \in M(\mathbb{C}) \\ j \neq i}} E_{\mathbb{Y}'_C}(j) < \sum_{\substack{j \in M(\mathbb{C}) \\ j \neq i}} E_{\mathbb{Y}_C}(j);$$

thus, $\exists j \in \mathbb{C} \ni E_{\mathbb{Y}'_C}(j) < E_{\mathbb{Y}_C}(j)$, from which it follows that $\mathbb{Y}$ is considerable. $\square$

Remark. The Nash equilibrium values for the different coalitions determine the value of each coalition, given the partition and the simplifying assumptions we have made. These simplifying assumptions, although quite limiting, are common in modern economics. Provided the Nash equilibrium is unique, this serves to define a partition value function for the game. Further assumptions would be needed to resolve the partition value function to a coalition function. The customary way of defining a coalition function, by adopting the minimum value that can be realized against the most hostile opposition, assumes that an agent will sometimes be taking active measures to reduce his own future payoffs, and so is not applicable in this case. Note that, since mixed strategies are among the strategies that define a considerable strategy array, the foregoing theorem assures us that a considerable strategy array exists. It does not assure us of uniqueness, but considerable strategy arrays are not unique in any case, given transferable utility.

8.5. Chapter Summary and Conclusions

This chapter has sketched, primarily through examples, an approach to a theory of intertemporal cooperative games. Of course, much remains to be done for a complete, rigorous theory, but the conclusion that can be drawn at this stage is that intertemporal games can consistently be expressed in partition function form. In doing so, however, it will be necessary to return to the concept of strategies proposed by von Neumann and Morgenstern. For intertemporal analysis, the underlying game needs to be expressed in extensive form, and a strategy understood as a complete plan of decisions contingent on all possible eventualities at each future period. For games in coalition function form, we assume that utility is transferable both from one agent to another and from one time period to another.

Chapter 9

A Theory of Enterprise

As we have seen, the Biform Game provides a coherent view of situations in which cooperative and noncooperative decisions are mixed. We have also seen that models of search and matching fit well in the Biform Game formalism. Moreover, such a model provides an answer to the question why, indeed, any decisions are made noncooperatively: people cannot cooperate until they are in contact with one another. Therefore, the process of establishing contacts is unavoidably noncooperative. Probably this deep insight will seem humdrum to people who are innocent of economics and game theory, especially to young people who spend much of their time "networking" and dating.

Search and matching models have been applied most extensively to employment relationships, but as Hall (2008) has observed, may also be applied to customer relationships. Such relationships are often persistent over time, as many employment relationships are. The expected persistence of the relationship will vary widely in both cases, but more persistent relationships of both kinds seem to be characteristic of the modern marketplace. It makes no more sense to model modern product markets as a bazaar or as an auction than it does to model modern employment relations as a day-labor market.

One of the simplifying assumptions adopted almost universally in cooperative game theory is that each agent is a member of one and only one coalition. Thus, the coalition structure can be expressed as a partition. In the actual world of economic activity, however, an individual will often be a member of more than one coalition: one or more as an employee or owner, one or more as a routine customer of a particular pharmacy or automobile service station, and still others for to procure other services and goods. Even a single, unrepeated act of exchange is the formation of a cooperative coalition, however brief. Thus, in place of a partition, the coalition structure would be an arbitrary set of subsets of the population. The model in the following section expands on that possibility.

9.1. An Overview of the Model

In this chapter, we revisit what economists call "the theory of the firm" from a point of view that treats "the firm" as a cooperative coalition among owners, employees, and customers, a coalition for production and sale. But this view incorporates also the element of noncooperative decisions in the formation of coalitions along lines suggested by the costly job matching models in recent macroeconomics. Thus, the chapter considers an enterprise economy as a mixture of cooperative and noncooperative decision-making and integrates those two aspects via the Biform Game formalism discussed in the previous two chapters. Thus, each period in the ongoing enterprise economy comprises two subperiods or stages, the first a noncooperative stage in which links are formed between prospective employers, employees, and customers. These links are formed via a costly probabilistic matching process. At the second stage, coalitions are formed. Coalitions can be formed only among those who have appropriate links formed at the first stage. Stable coalitions and imputations are limited by rationality constraints, as in a core solution; and within those limits, they will be determined as the NS-nucleolus of the coalitions formed.

Thus, the coalition structure, rationality constraints, and bargaining power of the agents will themselves be determined by the noncooperative decisions to undertake costly search for coalition matches at the first stage. The coalition structure, however, may not be a partition. Individuals may be members of different coalitions for different purposes, and may invest their resources in seeking links that would enable such membership.

These elements will require some preliminary formal development. This formal discussion can only be sketched, but hopefully the sketch will be sufficient to show that the Biform Game approach can coherently integrate the cooperative and noncooperative aspects of an enterprise economy, and the role that bargaining power must play in such an integration.

9.2. FCS Games

Let N be a set of agents $\{1, 2, \ldots, n\}$ and $S \subset N$. Let

$$\mathscr{S} = \{\{S_1, S_2, \ldots, S_m\} | S_i \subseteq N\},$$

m a positive integer. That is, $\mathscr{S}$ is a subset of the set of all sets of non-empty finite subsets of N. Each $\mathscr{Q} \in \mathscr{S}$ will be called a *feasible coalition structure* (FCS). Then $S \in \mathscr{Q} \in \mathscr{S}$ will be called an embedded coalition. In a special case of particular interest, $\mathscr{S}$ might be the set of all partitions of N.

For each $\mathscr{Q} \in \mathscr{S}$ there corresponds a set of strategies $\Sigma_{\mathscr{Q}} = \{\{\sigma_{i.k}\}_{i \in N}\}$. Let $\Sigma_i = \{\sigma_{i,k}\}$ be the set of all strategies under the control of Agent i. If $\forall \mathscr{Q} \in \mathscr{S}$ we have $\Sigma_{\mathscr{Q}} = \{\Sigma_i\}$ and $|\Sigma_i|$ finite, then the $S \in \mathscr{Q}$ would be coalitions in a game in normal form. Instead, we will focus on NTU games, so that the strategies for the coalitions in a coalition structure $\mathscr{Q}$ are the utility assignments for the members of $S \in \mathscr{Q}$. Further, for each $\mathscr{Q} \in \mathscr{S}$, to each $C \in \mathscr{Q}$, there corresponds a constraint function $f_{\mathscr{Q},C}$ as discussed in Chapters 3 (Section 3.3), 5, and 6 (Section 6.5), and the set of feasible coalitional strategies (utility assignments U_i) is determined by $f_{\mathscr{Q},C}(\{U_i\}_{i \in C}) \leq 0$.

In general the determination of the utility assignments for a particular coalition will reflect the bargaining powers of deviations from the coalition structure in which it is embedded. Corresponding to each $S \in \mathscr{Q} \in \mathscr{S}$, therefore, is a bargaining power vector ϕ_C. Note that it may not be possible to speak meaningfully of individual bargaining power, since it may be that there is no $\mathscr{Q} \in \mathscr{S}$ such that $\{i\} \in \mathscr{Q}$.

Accordingly, we identify a feasible condition structure game, FCS game, as a set

$$\Gamma = \{N, \mathscr{S}, \{f_{\mathscr{Q},S}(\{U_i\}_{i \in S}), \phi_S\} \; \forall \, S \in \mathscr{Q} \in \mathscr{S}\}. \tag{9.1}$$

For such a game, the cooperative-noncooperative distinction is modified, if not eliminated completely. Strategies are chosen by coalitions, not by individuals, and are specific to the coalition structures in which the coalitions are embedded. For a cooperative treatment, coalitions are supposed to be formed voluntarily, for mutual benefit, so that the expectation would be that the coalition will choose its strategy and payoffs by a consensus of the members. In such a case a concept of the core would seem to be applicable as a first step. As an alternative, we might identify an agent of a particular type, a "proprietor" or "entrepreneur," who has the power to determine the strategy and payoffs schedule of the coalition. This will be unambiguous if the feasible condition structures are comprised of coalitions that include exactly one member of that particular type. This could be seen as a noncooperative treatment. An alternate, more general noncooperative treatment might rank all agents and have the highest-ranking one make the decisions in each coalition. As an intermediate case, we might suppose that each coalition makes decisions by majority rule.

For a bargaining power solution to an FCS game, formally, each coalition will maximize its value at the distributional weights determined by the bargaining powers and the rationality constraints the coalition faces. We should stress that "value" in this sense has nothing necessarily to do with market values. Indeed, all three of the cases above — one person rule by the entrepreneur or person of highest rank, majority rule, and consensus — can be characterized by different distributions of bargaining power in a bargaining power solution.

Moreover, even in the "noncooperative" cases, if coalition membership is not compulsory, stability will require that the solution is within a set of core-like constraints. A case of compulsory membership — such as a state or feudal order — might be characterized by a FCS game with only a single feasible coalition structure or a very limited set of such structures.

For FCS games, as for games in partition function form, the results of a deviation from $\mathcal{Q} \in \mathcal{S}$ may be ambiguous. It will be necessary to restrict the concept of a deviation somewhat.

Definition 9.1. For an FCS game, a deviation from $\mathcal{Q} \in \mathcal{S}$ is a set $S \subseteq N \ni S \notin \mathcal{Q}, \exists \, \mathcal{R} \in \mathcal{S} \ni S \in \mathcal{R}$.

In case there is more than one such $\mathcal{R}$, we shall again rely on an arbitrary successor function $\mathcal{R} = \sigma(\mathcal{Q}, S)$. A bit more will be said about this in Section 9.4 of this chapter.

It remains to determine how the coalitions within a FCS interact to determine the values of the embedded coalitions. The assumption most widespread in the literature of cooperative game theory is the assurance principle: that each coalition is valued on the assumption that all other coalitions choose their most hostile strategies. However, (as previously argued in Chapter 8) this will hardly do for a cooperative treatment of an FCS game, in which strategies are chosen by consensus, since a single agent may be a member of more than one coalition. The assurance principle requires us to believe that the agent will support policies that most reduce his own payouts in his other coalitions. In the previous chapter, we faced the same possibility in that an individual agent might be a member of more than one coalition, at different times. For this possibility it was proposed that attention be limited to *considerable* presolutions. Among considerable presolutions, the set of solutions might be further limited by concepts such as the core or N-S nucleolus. It will be worthwhile explicitly to adapt this concept to FCS games. This will be addressed in the next subsection.

9.2.1. *Considerable Solutions in FCS Games*

Let Γ be an FCS game, $i \in N$, $\mathcal{Q} \in \mathcal{S}$, $\mathcal{Q} = \{C_1, \ldots, C_\xi, \ldots, C_m\}$, and let $U = \{U_{\xi,i}\}_{\substack{i \in C_\xi \\ C_\xi \in \mathcal{Q}_i}}$, U a feasible imputation for $\mathcal{Q}$; that is, $U_{\xi,i}$ is the benefit to Agent i from coalition ξ. As a first step we need to aggregate $U_{\xi,i}$ for Agent i. Following expression (23) in Chapter 8, Section 8.4, we might write

$$V_i = \sum_{C_\xi \in \mathcal{Q}_i} w_\xi U_{\xi,i} \tag{9.2a}$$

where w_ξ are constant weights. However, in Chapter 1, Section 1.3, it was argued that additivity would not be an appropriate assumption in a context such as this. Accordingly, instead, suppose that the overall well-being of Agent i derived from her membership in various coalitions is determined as

$$V_i = F_i(U_i) \quad \text{where } U_i = \{U_{\xi,i}\}_{C_\xi \in \mathcal{Q}_i} \tag{9.2b}$$

and F_i is continuously differentiable, concave and positively responsive, that is, given $U_i = \{U_{1,i}, \ldots, U_{\xi,i}, \ldots, U_{k,i}\}$, $U_i^* = \{U_{1,i}, \ldots, U_{\xi,i}^*, \ldots, U_{k,i}\}$, and $U_{\xi,i}^* > U_{\xi,i}$, then $F_i(U_i^*) > F_i(U_i)$. (That is, cet. par, larger payouts from any coalition are preferable to smaller.) Let U be an imputation for $\mathcal{Q}$. Then, since $U_i \subseteq U$, we may define $F_i(U) = F_i(U_i)$.

Definition 9.2. For an FCS game, then, U is considerable for $\mathcal{Q}$ if it is feasible for $\mathcal{Q}$, and for any feasible imputation $U^* \neq U$, $F_i(U^*) > F_i(U) \Rightarrow \exists j \in N \ni$

(1) $\exists C \in \mathcal{Q} \ni i \in C, j \in C$.
(2) $F_j(U^*) < F_j(U)$.

Remark. That is, U is Pareto-optimal within each coalition $C \in \mathcal{Q}$, taking account of payments to the members of C not only from C but also taking into account the payments they obtain from other coalitions in $\mathcal{Q}$.

Lemma 9.1. *If U is considerable for $\mathcal{Q}$, $C_\xi \in \mathcal{Q}$, $U_\xi = \{U_{i,\xi}\}_{i \in C_\xi}$, then U_ξ is Pareto-optimal among the members of C_ξ, for all feasible imputations for C_ξ.*

Proof. Suppose the contrary. Then $\exists U_\xi^* = \{U_{j,\xi}^*\}_{j \in C_\xi}$, a feasible imputation for $\mathcal{Q}$, and $i \in C_\xi \ni U_{i,\xi}^* > U_{i,\xi}$; moreover $\forall j \in C_\xi, j \neq i, U_{i,\xi}^* \geq U_{i,\xi}$.

$$\text{Construct } U^\dagger = \left\{ \begin{matrix} U_{j,\xi}^* & \text{for } j \in C_\xi \\ U_{j,\xi} & \text{for } j \in C_\zeta \in \mathcal{Q}, \ C_\zeta \neq C_\xi \end{matrix} \right\}.$$

Note that, from the feasibility of U^*_ξ, $f_{\mathscr{Q},C}(U^\dagger) \leq 0$, and from the feasibility of U, $f_{\mathscr{Q},S}(U^\dagger) \leq 0 \; \forall S \in \mathscr{Q}$, $S \neq C$. Therefore, $U^\dagger$ is feasible.

By the positive responsiveness of F_j, we will have

$$F_i(U^\dagger) > F_i(U), \quad F_j(U^\dagger) \geq F_j(U)$$

$\forall j \in C \backslash \{i\}$, contradicting the considerability of U. □

Remark. This proof relies on the assumption that $\frac{\partial f_{\mathscr{Q},C}}{\partial U_j} = 0$, $\forall j \notin C$. That is, it excludes a certain concept of externality. Otherwise, $U^\dagger$ might not be feasible, as the shift of C_ξ to U^*_ξ might shift the functions $f_{\mathscr{Q},S}$ for $S \in \mathscr{Q}$, $S \neq C$, in ways that would render $U^\dagger$ infeasible. Since this cooperative presolution will be considered as a second stage in a Biform Game, we suppose that the functions $f_{\mathscr{Q},S}$ are determined by noncooperative decisions at the first stage. For other applications, however, this model might need to be reconsidered.

Since U_ξ is Pareto-optimal for C_ξ, corresponding to U_ξ will be $\{\lambda_i\}_{i \in C_\xi}$, a set of distributive weights dual to U_ξ. Now consider a noncooperative game $\Gamma^\dagger$ in which the players are the coalitions $S \in \mathscr{Q}$ and the payoffs are

$$V_\xi = \sum_{i \in C_\xi} \lambda_i U_{i,\xi}. \tag{9.2c}$$

Remark. This does *not* parallel the constructed noncooperative game of Chapter 5, Section 5.2 and Chapter 6, Section 6.5, as only the coalitions in $\mathscr{Q}$ are players in the game, at this stage of the discussion.

Lemma 9.2. *Assume U is considerable for $\mathscr{Q}$ and U_ξ is as defined in Lemma 9.1. Then U_ξ is a best response in $\Gamma^\dagger$ so that U corresponds to a Nash equilibrium of $\Gamma^\dagger$.*

Proof. Assume the contrary. Then $\exists C_\xi \in \mathscr{Q}$ and a feasible imputation U^* for $\mathscr{Q} \ni \sum_{j \in C_\xi} \lambda_j U^*_{j,\xi} \geq \sum_{j \in C_\xi} \lambda_j U_{j,\xi}$. But this would contradict the definition of the λ_j as dual distributive weights in the maximization of $\sum_{j \in C_\xi} \lambda_j U_{j,\xi}$. □

Lemma 9.3. *Assume $U = \cup_{C_\xi \in \mathscr{Q}} U_\xi$ and the U_ξ are feasible for $\mathscr{Q}$ and constitute best responses in a Nash equilibrium of $\Gamma^\dagger$. Then U is considerable for $\mathscr{Q}$.*

Proof. Given distributional weights $\lambda_{i,\xi}$ for each $C_\xi \in \mathcal{Q}$, the Nash equilibrium of $\Gamma^\dagger$ can be characterized by

$$\max_{U_{i,\xi}} \sum_{i \in C_\xi} \lambda_{i,\xi} U_{i,\xi} \tag{9.2d}$$

subject to the feasibility constraint

$$f_{C_\xi, \mathcal{Q}}(U_\xi) \leq 0 \tag{9.2e}$$

for each C_ξ. From Definition 9.2 we may characterize a considerable imputation by a condition that, $\forall C_\xi \in \mathcal{Q}$

$$\max_{U_{j,\xi}} \sum \mu_{j,\xi} F_j(\{U_{j,\xi}\}_{C_\xi \in \mathcal{Q}}) \tag{9.2f}$$

subject, again, to (9.2e), where $\mu_{j,\xi}$ are the distributional weights that actually express power relationships within C_ξ. The Lagrangean function for this problem is

$$\mathcal{L} = \sum_{j \in C_\xi} \mu_{j,\xi} F_j(\{U_{j,\zeta}\}_{C_\zeta \in \mathcal{Q}_j}) - \theta f_{C_\xi, \mathcal{Q}}(\{U_{j,\xi}\}_{j \in C_\xi}). \tag{9.2g}$$

For $\zeta \neq \xi$, $U_{j,\zeta}$ is determined by the distributional weights $\mu_{j,\zeta}$ and the feasibility constraint $f_{C_\zeta, \mathcal{Q}}(\cdot)$, and so are given for the purposes of characterizing the Pareto optimum among the members of C_ξ. The necessary conditions from (9.2g) are

$$\mu_{j,\xi} \frac{\partial F_j}{\partial U_{j,\xi}} = \theta \frac{\partial f_{C_\xi, \mathcal{Q}}}{\partial U_{j,\xi}}. \tag{9.2h}$$

Setting

$$\lambda_{j,\xi} = \mu_{j,\xi} \frac{\partial F_j}{\partial U_{j,\xi}}. \tag{9.2i}$$

Equation (9.2h) will be identical to the necessary conditions for Eq. (9.2d). Since the sufficient conditions are identical, we may conclude that an imputation that satisfies (9.2d) with multipliers as determined by (9.2j) will also satisfy (9.2h) and so will be considerable. □

Theorem 9.1. *U is considerable iff it is a Nash equilibrium of the game* $\Gamma^\dagger$.

Proof. Lemma 9.3 establishes "if" and Lemmas 9.1, 9.2 establish "only if." □

9.2.2. The Core

In summary, a solution is considerable if the imputations are Pareto-optimal within each coalition and, with coalitions valued at appropriate distributional weights, the solution is a Nash equilibrium among the coalitions in each feasible condition structure. Let $\vartheta(\mathscr{Q})$ denote the set of all $\mathbf{U}$ that satisfy these conditions. The candidate solutions for a feasible condition structure game will be the pairs $\{\mathscr{Q}, \mathbf{U} | \mathbf{U} \in \vartheta(\mathscr{Q})\}$.

Lemma 9.4. *For any $\mathscr{Q} \in \mathscr{S}$, $\vartheta(\mathscr{Q})$ is non-null.*

This lemma will follow from the assumptions about the convexity of the constraint functions and from the application of Nash' theorem that a Nash equilibrium in pure strategies must always exist when the strategies are drawn from a closed compact set.

Now, let $\mathscr{Q} \in \mathscr{S}$, $S \notin \mathscr{Q}$, $S \in \mathscr{R} \in \mathscr{S}$, and $\mathscr{R} = \sigma(\mathscr{Q}, S)$. That is, S is a deviation from $\mathscr{Q}$ and $\mathscr{R}$ is its successor coalition structure. Let $\mathbf{U}_S = \{\mathbf{U}_i\}_{i \in S}$ be a considerable utility assignment for $i \in S$ that is feasible in coalition structure $\mathscr{Q}$ and suppose that $\mathscr{R} = \sigma(\mathscr{Q}, S)$ and $f_{\mathscr{R}, S}(U_S) \leq 0$. Then we may say that $\{\mathscr{Q}, \mathbf{U}\}$ is dominated via $S, \mathscr{R}$. (If = applies then tho domination io woak.) Now, consider

$$\Psi(\mathscr{Q}) = \{\mathbf{U} \text{ feasible for } \mathscr{Q} \mid \mathscr{R} = \sigma(\mathscr{Q}, S), \ f_{\mathscr{R}, S}(\mathbf{U}_S) > 0\}. \qquad (9.3)$$

Then $\Psi(\mathscr{Q})$ is the *coalition structure core* for $\mathscr{Q}$. Of course, $\Psi(\mathscr{Q}) = \varnothing$ is possible. Consider the set

$$\mathscr{D} = \{\mathscr{Q} \in \mathscr{S} | \Psi(\mathscr{Q}) \neq \varnothing\} \qquad (9.4)$$

$\mathscr{D}$ is the set of *stable coalition structures* for Γ. Then the core for Γ is

$$\Psi(\Gamma) = \{\{\mathscr{Q}, \mathbf{U}\} | \mathscr{Q} \in \mathscr{S}, \mathbf{U} \in \Psi(\mathscr{Q})\}. \qquad (9.5)$$

(Recall that $\mathbf{U} \in \Psi(\mathscr{Q}) \Rightarrow f_{\mathscr{Q}, C}(\mathbf{U}) \leq 0 \forall C \in \mathscr{Q}$.) The core $\Psi(\Gamma)$ and the set of stable coalition structures $\mathscr{D}$ clearly may be null sets, as the coalition structure core may be null for every $\mathscr{Q} \in \mathscr{S}$.

As before we will characterize the core imputation for each $\mathscr{Q}$ as the N-S nucleolus for $\mathscr{Q}$. We consider a Nash equilibrium among the embedded coalitions of all $\mathscr{Q} \in \mathscr{S}$ in which the strategies are the distributional weights λ_i and the utility assignments U_i and the payoffs are weighted sums of the gains of all deviations from the coalitions,

$$\sum_{S \in \Xi_C} \phi_S \ln \left[\sum_{i \in S} \lambda_i (U_i^{\mathcal{P}} - U_i^{\sigma(\mathcal{P}, S)}) \right] \qquad (9.6)$$

subject to the appropriate rationality constraints. For now, we will defer any further formal development. It may be noted, however, that this Nash equilibrium of all embedded coalitions entails both a Pareto optimum within each coalition and a Nash equilibrium among the coalitions in each coalition structure. From this it follows that the N-S nucleolus is considerable.

9.3. Determination of the Feasible Condition Structures

In the previous section, some concepts were sketched for a cooperative solution of games in which coalitions may be formed among the members of a set which is an element of an arbitrary set of subsets of the population, which may or may not constitute a partition. If we conceive enterprise as a process in which potential co-workers are first matched through a search and matching process that is at least partly random, this process determines the set of feasible condition structures.

We suppose that matching establishes links between players in the game *as individuals.* Thus the product of the matching process is a graph defined on the agents.[1] Let the graph be denoted by

$$\mathscr{H} = \{\{i, j\} | i \in N, j \in N\}. \tag{9.7}$$

Define

$$\mathscr{D}^0 = \{\mathscr{Q} = \{S_1, S_2, \ldots, S_r\} | \forall\, S_m \in \mathscr{Q},$$
$$\exists i \in S_m \ni j \in S_m, j \neq i \Rightarrow \{i, j\} \in \mathscr{H}\}. \tag{9.8a}$$

That is, a feasible coalition either is a singleton or comprises agents linked to a particular central Agent i who is also a member of the coalition. We may think of that agent as the entrepreneur for the potential coalition. This is, of course, not the only possible formulation, and we will explore some alternatives.

If instead

$$\mathscr{D}^* = \{\mathscr{Q} = \{S_1, S_2, \ldots, S_r\} | \ \forall S_m \in \mathscr{Q}, \ i \in S_m, j \in S_m \Rightarrow \{i, j\} \in \mathscr{H}\}. \tag{9.8b}$$

---

[1]For related models, see Myerson (1977), Haeringer (1999), Algaba, *et al.* (2001), van den Brink *et al.* (2007), Herings, P.J. *et al.* (2010). These papers are quite different in their objectives and details, but provide examples of models in which coalition formation is constrained by links in a graph, and despite the differences, this section was influenced in particular by Myerson.

This would require that the coalition comprise agents who are linked to one another. A feasible set defined in this way is a subset of $\mathscr{D}^0$, since the definition would be applicable only to a set in which every agent is central, and "entrepreneur." Yet again, the condition might be

$$\mathscr{D}^\dagger = \{\mathscr{Q} = \{S_1, S_2, \ldots, S_r\} | \forall S_m \in \mathscr{Q}, \; i \in S_m, \; j \in S_m, \; \exists k, l, \ldots, q \ni$$

$$\{i, k\} \in \mathscr{H}, \; \{k, l\} \in \mathscr{H}, \ldots, \{q, j\} \in \mathscr{H}, \text{ and } |k, l, \ldots, q| \leq v\}.$$

$$(9.8c)$$

Here, of course, v is a positive integer. Thus, a feasible coalition in $\mathscr{D}^\dagger$ would comprise agents with at most "v degrees of separation."

Consider Fig. 9.1. For this graph, any set of subsets would be feasible for $\mathscr{D}^\dagger$, assuming $v \geq 1$; and any set of two or fewer elements would be feasible for $\mathscr{D}^*$. For $\mathscr{D}^0$, any set of three or fewer agents would be feasible.

Consider Fig. 9.2. For $\mathscr{D}^\dagger$, with $v \geq 1$, as before, any subset is possible. For $\mathscr{D}^*$, as before, any subset with two or fewer members is possible, but none of three or more. The discussion of $\mathscr{D}^0$ will be more complex. Consider Table 9.1 which shows the feasible coalitions for $\mathscr{D}^0$ where each of the five agents is the central agent or "entrepreneur." Note the duplications in this list. There are eleven distinct feasible coalitions in this case; $\mathscr{S}$ will comprise any subset of this set of eleven coalitions.

Notice that $\mathscr{D}^0$ imposes no consistency requirement on coalitions that might exclude from some coalition structures coalitions that might be feasible for some other coalition structures. If, for example, we consider

Fig. 9.1. A simple graph.

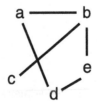

Fig. 9.2. A less simple graph.

feasible condition structures to be partitions, we are excluding as infeasible a coalition structure $\mathscr{Q}$ in which $C_1 \in \mathscr{Q}$, $C_2 \in \mathscr{Q}$, $C_1 \cap C_2 \neq \varnothing$; even though $\mathcal{P}_1 = \{C_1, N \backslash C_1\}$ and $\mathcal{P}_2 = \{C_2, N \backslash C_2\}$ would both be feasible, and so C_1 and C_2 are feasible in principle. A weaker limitation could be that an entrepreneur cannot be the central agent of more than one coalition. Using Table 10.1, this would in effect limit the set of feasible condition structures to those taking at most one coalition from each column. Using Table 9.1, for example, a coalition structure that would include $\{d, e\}$, $\{a, d, e\}$ and $\{b, d, e\}$ would be excluded, since it would require that either d or e be central to two coalitions at the same time. Thus we might define

$$\mathscr{D} = \{\mathscr{Q} = \{S_1, S_2, \ldots, S_r\} | \exists \nu : S_m \to N \ni$$
$$(1)\ \nu(S_m) \in S_m,$$
$$(2)\ i \in S_m \Rightarrow \{\nu(S_m), i\} \in \mathscr{H},$$
$$(3)\ S_m \in \mathscr{Q},\ S_n \in \mathscr{Q} \Rightarrow \nu(S_m) \neq \nu(S_n)\}. \tag{9.8d}$$

This rather complex condition seems appropriate to "a theory of enterprise." For example, using Table 9.1, we can see that there could be 72 feasible condition structures in which each of the five agents is an entrepreneur; and moreover, any subsets of those coalitions, with fewer than five coalitions, will be feasible condition structures.

One further elaboration will be explored. In the examples above there is only one sort of "membership" in the coalition. In an enterprise economy (or indeed in many other organizational schemas) memberships will be differentiated types — one may be a member as an employee, as a customer, or in other relationships. Links for the purpose of forming these relationships may themselves be distinct and impose different limits on the coalition structures that may be formed. For example, it is far more common for one agent to be a customer of a number of firms than to be employed by a number of firms. The employment relation is rivalrous or exclusive in a way that the customer relation is not.

Table 9.1. Coalitions feasible for $\mathscr{D}^0$ in Fig. 9.2.

a as central	b as central	c as central	d as central	e as central
$\{a, b\}$	$\{a, b\}$	$\{b, c\}$	$\{a, d\}$	$\{b, e\}$
$\{a, d\}$	$\{b, c\}$		$\{d, e\}$	$\{d, e\}$
$\{a, b, d\}$	$\{b, e\}$		$\{a, d, e\}$	$\{b, d, e\}$
	$\{a, b, c\}$			
	$\{a, b, e\}$			
	$\{a, b, c, e\}$			

Let N be a set of agents, and $\mathscr{E}$ and $\mathscr{H}$ be distinct graphs on N. Then

$$\mathscr{D}^{\#} = \{\mathscr{Q} = \{S_1, S_2, \ldots, S_r\}|$$

(1) $S_m = S_m^1 \cup S_m^2$

(2) $\exists i \ni$

(a) $j \in S_m^1 \Rightarrow$ either $j = i$ or $\{i, j\} \in \mathscr{H}$

(b) $j \in S_m^2 \Rightarrow$ either $j = i$ or $\{i, j\} \in \mathscr{E}$

(c) $S_m^1 \cap S_m^2 = \{i\}\}.$ $\qquad$ (9.8e)

This idea will be illustrated by a single example. See Fig. 9.3, in which the solid lines represent links that could correspond to employment relations and the dashed lines represent links that could form customer relations. Table 9.2. shows a few of the feasible coalitions.

Suppose both (1) and (3) are formed. Then $c, e,$ and f are at the same time customers of both b and d, presumably dividing their custom between them, depending on the offers the two may make to them. By contrast a can purchase only from b, because a has no customer link to d, although a has an employment link to d.

These examples are meant only to illustrate how existing links might constrain and determine the feasible condition structures for an FCS game.

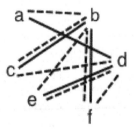

Fig. 9.3. Two superimposed graphs.

Table 9.2. Some feasible coalitions with Fig. 9.3.

	Central	Employees	Customers
(1)	b	c, f	a, c, e, f
(2)	h	c	a, f
(3)	d	a, e	c, e, f

A further formal analysis will not be advanced here. The graphs of links among agents will not be permanent in any case, but will be the product of an ongoing (costly) process of search for new links and matching of potential associates to enterprises. For the balance of the chapter we shift to a slightly more detailed discussion of the enterprise as a coalition for production and sale.

9.4. A Coalition for Production and Sale

In this chapter so far, we have explored the idea that the feasible coalitions are those among agents who are linked, where the links are formed through a noncooperative, and possibly costly, process of search for and matching to coalition partners. The objective for this section will be to make the idea more concrete by introducing some of the ideas often found in economic theories of "the firm," while placing these within the concepts of considerable imputations, the core, and the N-S nucleolus as a model of differentiated bargaining power. In a conventional economic model, agents play differentiated roles, as employees, proprietors, and customers of "the firm." Even where employer/employee relations are modeled as cooperative their relations with customers are usually modeled as noncooperative. (An important exception which has influenced this chapter is Hall, 2008). However, exchange is a cooperative act and, for a cooperative view of "the firm," customers should be considered as members of the coalition. This section will reflect that judgment.

A further complication will have to be dealt with. In practice the agents that participate in the noncooperative search and matching game will include already-existing coalitions, intending to recruit new employees and new customers, and, indeed, sometimes new capital partners. It is customary to assume that firms engaged in the search for new employees do so in such a way as to maximize profits, that is, without reference to the interests of "insider" employees and or of customers already associated with the firm. Now, if their rationality is bounded, or if attention is costly, it might be that the "insiders" and (especially) the customers might not bother to influence search-and-matching decisions. This might be subsumed to "rational inattention" (Sims, 2003). But this chapter is concerned with rational action theory and specifically with considerable solutions to games of production and sale, and for that purpose, coalitional decisions must be Pareto-optimal from the viewpoint of the current members of the coalition, in all three roles.

9.4.1. *A Production-and-Sale Coalition in a Steady State*

This chapter is concerned with strategies of production and sale and imputations of value that are stable in some sense. The core is one criterion of
stability. In language all economists are accustomed to, the core contains
arrangements that are stable against competitive pressures. However, the
core is timeless, as usually interpreted, and we might consider also stability
in the sense that the coalition maintains its scale of operation over time,
routinely replacing employees and customers who are lost, replacing capital
goods that are worn out, and so on. In macroeconomics and in the theory of
economic development, this is familiar as a stationary state, and models of
that sort play an important role in some labor market search-and-matching
models (e.g., Mortensen and Pissarides, 1994). This subsection will sketch
such a model.

We envision a future comprising periods $t = 1, 2, \ldots$, with each period
comprising two subperiods, a first subperiod in which a noncooperative
search and matching game is played and a second subperiod in which the
coalition plays out a cooperative game of production and exchange. At the
beginning of the first period the coalition has K_1 of undifferentiated capital
goods, m_1 employees, and n_1 customers. For a steady state these dimensions
will remain unchanged in later periods, but for now it will be useful to
retain the period subscripts. Each employee i exerts a variable effort $e_{i,t}$ in
period t. In period t, effective labor is

$$E_t = \sum_{i=1}^{n} e_{i,t}. \tag{9.9a}$$

Some of this effective labor is diverted to activities of searching for new
employees and customers. Denote by E_w the effective labor diverted to
searching for new employees and E_c that diverted to searching for new
customers, so that the remaining $E_t^* = E - E_w - E_c$ is directed to production. Production is assumed to be limited by the production function

$$Q_t = f(K_t, E_t^*) \tag{9.9b[2]}$$

---

[2]Thus we adopt the neoclassical production function in its most simplified, and arguably
oversimplified, form. We thus suppress such things as the heterogeneity of the capital
stock and business-to-business interdependency through the supply chain. This is done
largely for comparability and continuity with the orthodox theory of the firm. These more
realistic elements could be introduced into a FCS model, but they are largely dynamic
and so will make little difference for a stationary state model.

where K_t is the "capital stock" of the enterprise, supposed to be the property of the proprietor. For each employee the net benefit from participating in the coalition is

$$w_{i,t} - h_i(e_{i,t}) \tag{9.9c}$$

where $w_{i,t}$ is a side payment to the employee from the coalition and h_i is a disutility or effort-aversion function for Agent i that expresses the minimum money payment necessary to compensate him for the effort required. For each customer, we suppose the net benefit from participating in the coalition is

$$g_j(q_{j,t}) + M_{j,t} - S_{j,t} \tag{9.9d}$$

with

$$\sum_{j=1}^{n} q_{j,t} \le Q_t \tag{9.9e}$$

where g_j is Agent j's utility function for the good sold by the coalition, expressed in monetary terms, $M_{j,t}$ is the agent's spending on other goods, and $S_{j,t}$ is a side payment from the consumer to the coalition.

For simplicity we assume that there is only a single proprietor. The proprietor's profit is

$$\sum_{j=1}^{n} S_{j,t} - \sum_{i=1}^{m} w_{i,t} - F_t \tag{9.9f}$$

where F_t is real depreciation of the capital of the enterprise. But profit (in a particular period) is not the same as the proprietor's payoff (in the particular period). The proprietor will set aside some part of the profit to be reinvested, to offset the real depreciation of capital. This reinvestment will yield payoffs only in subsequent periods. Denoting plowback investment in period t by R_t, the proprietor's payoff in period t is

$$\sum_{j=1}^{n} S_{j,t} - \sum_{i=1}^{m} w_{i,t} - R_t. \tag{9.9g}$$

For a considerable solution, it is necessary both that the strategy and imputation be Pareto-optimal for the coalition in each period and that the strategies and imputations correspond to a Nash equilibrium among the coalitions at each stage. The flows E_w, E_c, and R_t determine the dimensions of the coalition at subsequent stages, probabilistically in the case of E_w and E_c, and perhaps R_t as well. For present purposes, since our concern with be with a steady state, and in what follows certainty-equivalence will be *assumed* but not modeled. Further, some of the members of the coalition in period 1 will not be members in later periods. We will adopt the customary radioactive-decay assumption (e.g., Hall, 2005) and let the probability of a separation for employee i between periods t and $t+1$ be a constant v. (We might think of such a separation as the evaporation of one of the links in the underlying graph discussed in the previous section.) Thus, in the absence of new recruitment, the work force in period $t+1$ would be $(1-v)m_t$. Let the number of new recruits be a monotonically increasing function ϕ of E_w, so that

$$m_t = (1-v)m_{t-1} + \phi(E_{w,t-1}). \tag{9.10a}$$

It follows moreover that the expected payoff to worker i from the coalition over the indefinite future is

$$\Pi_{w,i} = \sum_{\tau=t}^{\infty} \delta^{\tau-1}(1-v)^{\tau-1}(w_{i,\tau} - h_i(e_{i,\tau})). \tag{9.11a}$$

Similarly we suppose that customers drop out of the coalition at a radioactive-decay rate u, and new customers arrive according to a function $\psi(E_{c,t})$, so that

$$n_t = (1-u)n_{t-1} + \psi(E_{c,t-1}) \tag{9.10b}$$

and the payoff to customer j over indefinite future time is

$$\Pi_{c,t} = \sum_{\tau=t}^{\infty} \delta^{\tau-1}(1-u)^{\tau-1}(g_j(q_{j,t}) + M_{j,t} - S_{j,t}). \tag{9.11b}$$

Similarly, reinvested capital will be offset by depreciation, so that

$$K_t = K_{t-1} - F_{t-1} + R_{t-1}. \tag{9.10c}$$

Here again, for simplicity and comparability with a wide literature, we adopt a radioactive-decay model of depreciation of capital goods, with the rate of depreciation denoted by γ. Thus, $F = \gamma K$.

The payoff to the proprietor over indefinite future time will reflect the fact that firms do not last forever. If the firm were to be discontinued or bankrupted, the proprietor would be "separated" from the coalition. Such a separation could also come about with the sale or bequest of the proprietor's property rights. Once again, for simplicity we adopt a radioactive decay rule for this sort of separation, so that the probability that the owner will not continue as such for the coming period is a constant z. The proprietor's payoff for the indefinite future is

$$\Pi_{p,t} = \sum_{\tau=t}^{\infty} \delta^{\tau-1}(1-z)^{\tau-1} \left(\sum_{j=1}^{n} S_j - \sum_{i=1}^{m} w_i - R_i \right). \qquad (9.11c)$$

To further simplify the model, we will impose a "steady state," so that, for $t \neq \tau$,

$$K_t = K_\tau \quad E_{w,t} = E_{w,\tau} \quad E_{c,t} = E_{c,\tau}$$
$$w_{i,t} = w_{i,\tau} \quad q_{j,t} = q_{j,\tau} \quad S_{j,t} = S_{j,\tau} \qquad (9.12a)$$
$$e_{i,t} = e_{i,\tau} \quad M_{j,t} = M_{j,\tau} \quad m_t = m_\tau$$
$$n_t = n_\tau.$$

Accordingly, time subscripts will be suppressed from this point on. These assumptions imply that

$$m = \frac{1}{v}\phi(E_w) \qquad (9.12b)$$

$$n = \frac{1}{u}\psi(E_c) \qquad (9.12c)$$

$$K = \frac{1}{\gamma}R. \qquad (9.12d)$$

9.4.2. *Optimality within the Coalition*

In these simplified terms, from the perspective of the first period, the discounted expected values of payoffs will be

$$\Pi_{w,i} = A(w_i - h_i(e_i)) \qquad (9.13a)$$
$$\Pi_{c,j} = B(g_j(q_j) + M_j - S_j) \qquad (9.13b)$$

$$\Pi_p = C \left(\sum_{j=1}^{n} S_j - \sum_{i=1}^{m} w_i - R \right). \qquad (9.13c)$$

In all probability we will not have $A = B = C$. Plausibly $u > v > z$, so that $C > A > B$. Nevertheless, intertemporal side payments will not be considered. We may characterize Pareto optimum and the utility-possibility frontier within the coalition in the first period by

$$\max V = \sum_{i=1}^{m} \lambda_i (w_i - h_i(e_i)) + \sum_{j=1}^{n} \lambda_j (g_j(q_j) + M_j - S_j)$$

$$+ \lambda \left(\sum_{j=1}^{n} S_j - \sum_{i=1}^{m} w_i - \gamma K \right). \tag{9.14a}$$

Here, the adjusted discount rates, A, B, C are subsumed to the distributive weights, $\lambda_i, \lambda_j, \lambda$. The maximization is subject to the constraints (9.8a), (9.8b), and (9.11b)–(9.11d). Note, however, that (9.11b) and (9.11c) are integer constraints. For the steady-state model of a coalition with a fixed membership, we instead impose the equivalent target constraints

$$E_w \geq E_w^{\dagger} = v\phi^{-1}(m) \tag{9.12e}$$

$$E_c \geq E_c^{\dagger} = u\psi^{-1}(n). \tag{9.12f}$$

Thus we form the Lagrangean function

$$\mathcal{L} = V + \mu_1 \left(\sum_{i=1}^{m} e_i - E_w - E_c - E^* \right) + \mu_2 \left(f(K, E^*) - \sum_{j=1}^{n} q_j \right)$$

$$+ \mu_3 (E_w - E_w^{\dagger}) + \mu_4 (E_c - E_c^{\dagger}) + \mu_5 (R - \gamma K). \tag{9.14b}$$

Because many of these Lagrange multipliers enter into the determination of side payments, this plethora of Lagrange multipliers collapses into a few. In particular, $\lambda_i = \lambda_j = \lambda$. Thus, the maximand V is proportional to

$$\max V^* = \sum_{j=1}^{n} [g_j(q_j) + M_j] - \sum_{i=1}^{m} h_i(e_i) - \gamma K. \tag{9.14c}$$

In ordinary language terms, the Pareto-optimality requires that the coalition maximize the sum of producers' and consumers' surpluses for its members and may then distribute that value by side payments in any way that is consistent with rationality constraints. In more game-theoretic terms, the value of a coalition is well-defined and is (in a sense) a market value.

Moreover, we derive the familiar efficiency conditions

$$\frac{\partial g_j}{\partial q_j} = \frac{\partial g_k}{\partial q_k}, \ \forall j, k \tag{9.15a}$$

$$\frac{\partial h_i}{\partial e_i} = \frac{\partial h_k}{\partial e_k}, \ \forall i, k \tag{9.15b}$$

$$\frac{\partial g_j}{\partial q_j} \frac{\partial f}{\partial E^*} = \frac{\partial h_i}{\partial e_i} \tag{9.15c}$$

and the less familiar

$$\frac{\partial g_j}{\partial q_j} \frac{\partial f}{\partial K} = \gamma. \tag{9.15d}$$

We note that the rate of discount does not appear in (9.14d), but recall that discounting of future payments to present values is already subsumed in λ. Moreover, the payoff to the proprietor is gross of second-best alternative costs, which would be introduced via a rationality constraint. If the proprietor chooses to spend his capital "playing at ducks and drakes" rather than reinvesting it, that is a deviation from the coalition. In addition we have

$$\mu_3 = \lambda \frac{\partial h_i}{\partial e_i}. \tag{9.15e}$$

That is, the marginal value of additional spending on labor recruitment is proportional to the value of the marginal disutility of labor. That is, *with a given number of customers*, the benefit of recruiting more labor is to decrease the work-load per employee. Further

$$\mu_4 = \lambda \frac{\partial g_j}{\partial q_j} \frac{\partial f}{\partial E^*}. \tag{9.15f}$$

That is, the marginal value of spending on recruiting new customers is proportional to value of the marginal product of labor. That is, *with a given labor force*, the value of another customer can be measured by the influence of the additional customer on the value productivity of labor in the coalition. [Equation (9.14c) tells us that these conditions are equivalent but the presentation in (9.14e) and (9.14f) suggest the natural interpretations of the conditions.]

9.4.3. *Bargaining Power*

We will explore the results of the following bargaining power hypothesis:

(1) Employee i's best alternative payoff, that is, his "reservation wage," is w_i^*; that is, his rationality constraint is $w_i - h_i(e_i) \geq w_i^*$.

(2) Customer j can obtain a net benefit of u_j^* without participating in coalition C, either by giving up the good sold by the coalition or by re-affiliating with another seller. Thus j's individual rationality constraint is $g_j(q_j) + M_j - S_j \geq u_j$, or, equivalently, $S_j \leq g_j(q_j) + M_j - u_j$.

(3) The proprietor as a singleton has bargaining power $\phi_j = 1$. (This is a normalizing assumption.) The proprietor's rationality constraint is $\Pi_p \geq \rho K$.

(4) Individual customers each have identical bargaining power ϕ_c.

(5) Individual employees have no bargaining power, i.e., $\phi_i = 0$.

(6) The employees as a group have positive bargaining power $\phi_w > 0$.

(7) In the case of a bargaining impasse in period t, customers obtain their rationality constraints, employees and proprietors obtain zero in the current period.

(8) All other deviations have zero bargaining power.

For employees and the proprietor, the bargaining "threat point" differs from the rationality constraint. A bargaining impasse *in the current period* does not imply the dissolution of the coalition, which may resume, with the usual turnover of labor and customers, in the following period. In any case, when we consider the work group as a deviation from coalition C, its rationality constraint is distinct from that of its individual members, and corresponds to the value that the work group could realize separately from the other members of the coalition, and zero seems a reasonable estimate, in the absence of any special conditions. The clearest instance is the threat of a strike or a lock-out. There may be other threats — such as a "slow-down," that is, effort withdrawal, or a unionization campaign where a union is absent — that would result in a reduced, but still positive payoff for the proprietor and for the employees, but this is ignored for simplicity. If we consider a singleton coalition of an individual employee as a deviation from coalition C, that is, a quit, the individual may have links that will enable him immediately to reaffiliate with another employer. If not, he will become "unemployed" and await another round of link formation at the first-subperiod of the following period. His individual rationality constraint will reflect this. Customers, however, are assumed to have no threat strategies as a group (boycotts would be an exception) and are more likely to have links to other sellers of substitutable goods that would enable them to maintain their consumption levels nearly undiminished in the context of a strike, lockout, or other suspension of production in the current period.

Accordingly, set

$$A^* = \sum_{t=2}^{\infty} \delta^{t-1}(1-v)^{t-1}(w_{i,t} - h_i(e_{i,t})) \tag{9.16a}$$

$$B^* = \sum_{t=2}^{\infty} \delta^{t-1}(1-u)^{t-1}(g_j(q_{j,t}) + M_{j,t} - S_{j.t}) \tag{9.16b}$$

$$C^* = \sum_{t=2}^{\infty} \delta^{t-1}(1-z)^{t-1}\left(\sum_{j=1}^{n} S_{j,t} - \sum_{i=1}^{m} w_i - R_t\right). \tag{9.16c}$$

For a stationary state, from the view of the first period, A^*, B^*, and C^* will be constants. Accordingly, we characterize a N-S nucleolus by

$$\max V^\dagger = \phi_w \ln \sum_{i=1}^{m} \lambda_i[(w_{i,1} - h_i(e_{i,1})) + A^*]$$

$$+ \sum_{j=1}^{n} \phi_j \ln \lambda_j[g_j(q_{j,1}) + M_{j,1} - S_{j,1} - u_j + B^*]$$

$$+ \ln \lambda \left[\sum_{j=1}^{n} S_{j,1} - \sum_{i=1}^{m} w_{j,1} - R + C^*\right] \tag{9.17a}$$

subject to appropriate rationality, normalization and feasibility constraints. For now, rationality constraints other than those for individual employees and consumers and the proprietor will be ignored, i.e., assumed non-binding. It will be convenient to compound the dynamic constraint $R \geq \gamma K$ with the proprietor's rationality constraint. Accordingly, the Lagrangean function will be

$$\mathcal{L} = V^{\dagger} + \mu_1 \left(\sum_{i=1}^{m} e_i - E_w - E_c - E^* \right) + \mu_2 \left(f(K, E^*) - \sum_{j=1}^{n} q_j \right)$$

$$+ \mu_3(E_w - E_w^{\dagger}) + \mu_4(E_c - E_c^{\dagger}) + \nu \left(1 - \sum_{i=1}^{m} \lambda_i - \sum_{j=1}^{n} \lambda_j - \lambda \right)$$

$$+ \sum_{i=1}^{m} \nu_i(w_{i,1} - h_i(e_{i,1}) - w_i^*) + \sum_{j=1}^{n} \nu_j(g_j(q_{j,1}) + M_{j,1} - S_{j,1} - u_j)$$

$$+ \nu_p \left[\sum_{j=1}^{n} S_{j,1} + \sum_{i=1}^{m} w_{i,1} + C^* - (\rho + \gamma) \right]. \qquad (9.17b)$$

Setting

$$\Pi = \sum_{j=1}^{n} S_{j,1} - \sum_{i=1}^{m} w_{j,1} - R + C^* \qquad (9.17c)$$

and

$$\Omega = \sum_{i=1}^{m} \lambda_i[(w_{i,1} - h_i(e_{i,1})) + A^*] \qquad (9.17d)$$

$$\Theta_j = g_j(q_{j,1}) + M_{j,1} - S_{j,1} - u_j + B^*. \qquad (9.17e)$$

We have, among the necessary conditions,

$$\frac{\partial \mathcal{L}}{\partial \lambda_i} = \frac{\phi_w}{\Omega}[(w_{i,1} - h_i(e_{i,1})) + A^*] - \nu \leq 0 \qquad (9.18a)$$

$$\frac{\partial \mathcal{L}}{\partial \lambda_j} = \frac{\phi_j}{\lambda_j} - \nu \leq 0 \qquad (9.18b)$$

$$\frac{\partial \mathcal{L}}{\partial \lambda} = \frac{1}{\lambda} - \nu \leq 0. \qquad (9.18c)$$

We may assume interior solutions in these cases as a condition of the Pareto optimality of the solution as among the members of the coalition. These conditions yield

$$\lambda_j = \phi_j \lambda. \qquad (9.19a)$$

That is, the distributional weight for a customer is proportional to the customer's individual bargaining power. This seemingly common-sense result reflects the assumption that customers have no collective threats, so that all sets of two or more customers are weighted at zero in the maximization. The case for employees is more complex. We have

$$\frac{\phi_w[(w_{i,1} - h_i(e_{i,1})) + A^*]}{\Omega} = \frac{1}{\lambda}. \tag{9.19b}$$

A further necessary condition for the maximum is

$$\frac{\partial \mathcal{L}}{\partial w_{i,1}} = \frac{\phi_w}{\Omega}\lambda_i - \frac{1}{\lambda\Pi} + \nu_i - \nu_p \le 0. \tag{9.18d}$$

Supposing that w_i^* is positive, an interior solution will be necessary for each i. Note that $\nu_i > 0$ only if the rationality constraint for i is binding. In that case w_i is constraint-determined. This would be likely in the case of individual employees with relatively high reservation wages, i.e., excellent competitive alternatives. In general

$$\lambda_i = \frac{\Omega}{\Pi}\frac{1}{\lambda\phi_w} + (\nu_p - \nu_i)\frac{\Omega}{\phi_w}. \tag{9.19c}$$

Suppose that for i, the rationality constraint is non-binding. Then

$$\lambda_i = \frac{\Omega}{\phi_w}\left(\frac{1}{\Pi} + \nu_p\right). \tag{9.19d}$$

Since the RHS comprises constants, λ_i is a constant for all i in this category. Put otherwise, the bargaining power of the work group puts a minimum under the distributive weights of the employees.

Further,

$$\frac{\partial \mathcal{L}}{\partial S_{j,1}} = -\frac{\phi_j}{\Theta_j} + \frac{1}{\Pi} - \nu_j + \nu_p \le 0. \tag{9.18e}$$

Again, if $\nu_j > 0$, then S_j is constraint-determined. In general,

$$\Theta_j = \frac{\phi_j\Pi}{1 + \Pi(\nu_p - \nu_i)}. \tag{9.19e}$$

However, if the rationality constraint for customer j is not binding,

$$\Theta_j = \frac{\phi_j\Pi}{1 + \Pi\nu_p}. \tag{9.19f}$$

That is, the consumer's surplus retained by the customer is no less than proportional to the individual customer's bargaining power, but may be greater if the customer's rationality constraint is binding.

Further yet again,

$$\frac{\partial \mathcal{L}}{\partial e_{i,1}} = -\frac{\phi_w}{\Omega} \lambda_i \frac{\partial h_i}{\partial e_{i,1}} + \mu_1 - \nu_i \frac{\partial h_i}{\partial e_{i,1}} \leq 0 \qquad (9.18\text{f})$$

$$\frac{\partial \mathcal{L}}{\partial q_{j,1}} = \frac{\phi_j}{\Theta_j} \frac{\partial g_j}{\partial q_{j,1}} - \mu_2 + \nu_j \frac{\partial g_j}{\partial q_{j,1}} \leq 0 \qquad (9.18\text{g})$$

$$\frac{\partial \mathcal{L}}{\partial E^*} = -\mu_1 + \mu_2 \frac{\partial f}{\partial E^*} \leq 0. \qquad (9.18\text{h})$$

For (9.18h) it seems appropriate to assume an interior solution since $E^* = 0$ would be a zero effective labor input situation, which on the usual assumptions would mean no output. For (9.18f) and (9.18g), corner solutions cannot be excluded in principle, but they would mean respectively that no effort is required of Agent i or that no output is allocated to Agent j. In those cases, Agent i or j would presumably be expelled from the coalition, and a coalition with zero output presumably would dissolve. Since we are concerned here with allocations in stable coalitions, these corner solutions will be ignored. Therefore, we have

$$\left[\frac{\phi_j}{\Theta_j} + \nu_j \right] \frac{\partial g_j}{\partial q_{j,1}} \frac{\partial f}{\partial E^*} = \left[\frac{\phi_w}{\Omega} \lambda_i + \nu_i \right] \frac{\partial h_i}{\partial e_{i,1}}. \qquad (9.20\text{a})$$

The left-hand side is a (weighted) value of the marginal product of effective labor, and the right-hand side is a (weighted) marginal disutility of labor. If, further, there is an employee i and a customer j such that $\nu_i = \nu_j = 0$, i.e., the individual rationality constraints are not binding, then, noting that

$$\frac{\phi_w}{\Omega} \lambda_i = \frac{1}{\Pi} + \nu_p \qquad (9.20\text{b})$$

$$\frac{\phi_j}{\Theta_j} = \frac{1}{\Pi} + \nu_p \qquad (9.20\text{c})$$

we have

$$\frac{\partial g_j}{\partial q_{j,1}} \frac{\partial f}{\partial E^*} = \frac{\partial h_i}{\partial e_{i,1}}. \qquad (9.20\text{d})$$

That is, the (unweighted) value of the marginal product of effective labor is equal to the marginal disutility of labor. Moreover, the output and effort are allocated so that the marginal utility of output is the same for each customer and the marginal disutility of effort is the same for each employee. This familiar efficiency condition may be violated for an individual, however, if (for example) it is cheaper to retain an employee whose individual rationality constraint is binding by reducing his work-load rather than by increasing his wage. Further, in this case we recover Eqs. (9.14e) and (9.14f) from the previous subsection, with the interpretation given there.

Finally,

$$\frac{\partial \mathcal{L}}{\partial K} = \mu_2 \frac{\partial f}{\partial K} - \nu_p(\rho + \gamma) \leq 0. \tag{9.18j}$$

That is, with $K > 0$,

$$\left(\frac{\phi_j}{\Theta_j} + \nu_j\right) \frac{\partial g_j}{\partial q_{j,1}} \frac{\partial f}{\partial K} = \nu_p(\rho + \gamma). \tag{9.21a}$$

If, then, the proprietor's rationality constraint is nonbinding, so that $\nu_p = 0$, then

$$\frac{\partial g_j}{\partial q_{j,1}} \frac{\partial f}{\partial K} = 0. \tag{9.21b}$$

However, with conventional assumptions on consumer preferences and substitutability of inputs, this cannot occur. Thus we conclude that the proprietor will increase the capital committed to the coalition up to the point that her rationality constraint is binding.

In this discussion, the bargain between the enterprise and a customer is modeled as an all-or-nothing offer, rather than a fixed-price offer. This is the natural translation of "side payment" from game theory to economics. Further, if the enterprise has some monopoly power, as it always will have in a world of costly matching of customers with sellers, a fixed-price offer will result in some foregone gains from trade, conditions that are excluded by the assumption that the relationship of the enterprise and its customers is (essentially) cooperative, and that would be excluded as well by the famous Barro (1977) critique in macroeconomics. In the actual world all-or-nothing offers are uncommon but price discrimination is as common as dirt. The cooperative perspective provides an explanation for the widespread observation of price discrimination. We may suppose that in practice, price discrimination approximates the distribution of benefits to consumers that would be realized by an ideal all-or-nothing bargaining framework.

Thus, for customer j, the "marginal price," $\frac{\partial g_j}{\partial q_{i,1}}$ will not in general be equal to the "average price," $\frac{S_j}{q_{j,1}}$. Suppose, however, that they *are* equal so that $S_{j,1} = q_{j,1}\frac{\partial g_j}{\partial q_{j,1}}$ and moreover $\nu_j = 0$. From (9.17e), we have

$$\phi_j = \frac{g_j(q_{j,1}) + M_{j,1} - q_{j,1}\frac{\partial g_j}{\partial q_{j,1}} - u_j + B^*}{\lambda\Pi}. \tag{9.22}$$

That is, to identify the average and the marginal price is to assume a quite specific, nonzero bargaining power of the customer.

9.5. Competition and Monopoly

As we have seen, once the coalition for production and sale has been formed, the coalition will have some "monopoly power" *vis a vis* the customers. However, despite a century of monopoly theory in neoclassical economics, there is no reason to expect "monopoly restriction" of output. The Marshall–Lerner conditions, that output corresponds to the equality of marginal revenue with marginal cost, reflect two assumptions: first, that the law of one price applies in the presence of monopoly, and second, that the price is determined so as to maximize profit, that is, noncooperatively. Clearly the latter assumption is out of place in a cooperative model. The first assumption, the law of one price, is deducible from a model of large group (noncooperative) competition in the absence of costs of transaction or search. As George Stigler (1961) observed in the paper that founded the search cost literature, a degenerate distribution of prices in inconsistent with costly search. Even if the relationship between the seller and the customer is noncooperative, it is implausible that the seller, having invested in recruiting the customer, would then discourage him from purchasing an efficient amount by making a fixed-price offer. Rather, a rational, noncooperative monopolist would establish price incentives, such as member privileges and volume discounts, to encourage an efficient rate of purchase.

Of course, where a law of one price is established despite monopoly conditions, by regulation or other institutional circumstances, the Marshall–Lerner conditions (or some more complex conditions) might apply. One "institutional circumstance" that may enforce a law of one price is a competitive secondary market for the monopoly's product (McCain, 2009, pp. 142–147; McCain, 1987). An illustrative example is found in the "scalping" of tickets for performances and sporting events. Every performance or sporting event is a differentiated product, so the markets for tickets are monopoly markets. Here, sellers have advocated "anti-scalping laws" to make the secondary market illegal, so that the monopolist can retain a price below the market value of a popular performance. The difficult question is why (in a noncooperative view of the monopoly) the monopoly would wish to do this. McCain (1987) suggests a somewhat complicated noncooperative hypothesis to explain it. In the light of a cooperative model of the monopolist, it is less mysterious. Keeping a low price for popular events, with non-price rationing, is in effect price discrimination in favor of loyal customers, who will also attend many less popular offerings, against casual customers, who are interested only in the popular event. It would function somewhat like a volume discount.

"Loyal" customers may have greater bargaining power. In any case, in a model of costly recruitment of customers, the loyal customer will provide a better return on the investment in recruiting him. Thus, even in a noncooperative model with matching costs for customers, there is reason to favor the "loyal" customer. In the model in the previous section, we have followed the crowd (and simplified the problem greatly) by assuming that separations of customers and workers occur according to a radioactive decay rule, with no difference from one agent to another. For many businesses, however, there will be casual as well as long-term associates among their customers, and the latter are likely to have more bargaining power reflecting their greater value to the coalition. The tendency of sellers to favor "loyal" customers is not new, and in the extreme conditions of wartime scarcity and rationing, casual customers may have been rejected completely in order to favor the loyal ones (Lewis, 1942). This tendency seems quite natural in a cooperative framework, but presents explanatory problems in a noncooperative view.

We have observed that "monopoly restriction" of output will not occur in the cooperative stage of our model. This is not to say the monopoly will have no effect on the allocation of resources. In the first instance monopoly power will reduce an individual customer's distributive weight λ_j by reducing the rationality constraint u_j^*. In an ideal case of pure monopoly, the customer's only alternative to buying from the monopolist is to entirely abstain from the product the monopolist sells, i.e., $u_j^* = f_j(0) + M_j$.

(This is true whether the monopoly power exists because there are no other suppliers or because the customer has no links to any.) Monopoly power may also reduce the customer's bargaining power. Recall that in a cooperative framework bargaining power arises from threats that the agent might carry out despite the fact that they "will not be something [the bargainer] would want to do, just for itself." Thus, if an individual shifts to his second-best alternative and thus is better off, this is expressed by the individual rationality constraint. But even if the shift would make the customer worse off, he may threaten it if the seller's terms are not satisfactory to him. This is what we mean by bargaining power, and monopoly, by depriving the customer of alternatives he might otherwise have, will reduce his bargaining power. Both by reducing the rationality constraint and by reducing the customer's bargaining power, monopoly would tend to reduce the distributive weight λ_i and thus the consumer's surplus retained by the customer, Θ_j.

But these results refer to the second, cooperative sub-period in the intertemporal game of production and sale. The *expected* outcomes at this stage will affect the first sub-period of noncooperative search and matching. Monopoly will play an important role here too. In general, the potential customer will balance the cost of searching for a supplier against the consumer's surplus he can expect to gain if the search is successful, considering also the probability of success. Suppose then that the monopolist will reduce the consumer's surplus nearly to zero. Then the potential customer may simply choose not to search, dropping out of the market. If potential customers differ in the consumer's surpluses that they may reasonably expect, perhaps because of differences in taste, then some may decline to search, dropping out of the market, who would search and transact if more competition resulted in better terms from the sellers. This is more or less the monopoly restriction of output observed in the Marshall–Lerner model. Since it is a noncooperative phenomenon, it is not surprising to find the restriction of monopoly output occurring at the noncooperative, search and matching stage of the game. Even this may not be the whole story. Recognizing that poor terms for customers associated with the firm may discourage customers from entering the market in the future, the production-and-sale coalition might find it best to ameliorate the offer to current customers. We might formalize this by making $E_c^{\dagger} = u\psi^{-1}(n, \sum_{j=1}^{n} S_j)$, with $\frac{\partial \psi^{-1}}{\partial \sum_{j=1}^{n} S_j} > 0$; that is, an increase in the sum of customer side payments would increase the cost of retaining a stable number of customers. The formal results of such a change will not be explored here, however.

This strategy of improving the routine offer to customers in order to encourage new customers to enter the market is especially likely for a monopoly. Suppose instead that several firms search for customers from the same pool. Then a reduction of S_j by one firm, with a view to encouraging the entry of more potential customers into the pool, would generate a positive externality to the other firms. Such an initiative would be unlikely in this more "competitive" situation. (We might, however, observe industry associations to promote entry into the common pool of potential customers by advertising or even by measures to encourage all competitors to make better offers to their customers.)

What has been said about monopoly power in consumer markets applies also to monopsony power in labor markets, where it will be less novel (e.g., Burdett and Mortensen, 1998).

For a model of costly search and matching of sellers and customers, monopolistic competition will be the usual case. "Product differentiation" will appear in a quite different light. Where the characteristics of the product are not dictated by some technological institutional or customary considerations, they will be chosen cooperatively by the production-and-sale coalition. That is, in more ordinary language, the firm will adjust the characteristics of its product primarily with a view to satisfying its customers preferences, not with a view to reducing its elasticity of demand and so raising the price consistent with the Marshall–Lerner conditions.

For a long-run perspective, competitive conditions are also characterized by "free entry," and that in turn leads to a condition of zero (economic) profit. In this model, the proprietor's second-best opportunity is expressed as her rationality constraint. Thus, the equivalent of the zero-economic-profit condition is that the proprietor's rationality constraint is binding. We already have that condition as a result of Eq. (9.20a), a necessary efficiency condition within the coalition without any assumption about entry. (But the assumed second-best constraint assumes away any increase of capital cost as a result of leverage, and a more realistic assumption at this point might modify the results.

In this chapter, the successor function has been taken as given. The successor function is a makeshift designed to incorporated in abstract the following sort of argument: a group S might consider deviating from coalition structure $\mathscr{Q}$, but anticipate that such a deviation could result in a further reorganization of $N \backslash S$ that would make some of the members of S worse off, and so restrain themselves and not deviate. Such a deviation would realize an opportunity for some profit or surplus for group S, but that surplus would in turn be disrupted by the subsequent reorganization. In an enterprise economy, however (unlike a multiparty parliament as in the example in McCain, 2009, p. 180) such an opportunity is likely to be available to other groups beside S. The members of S may reason that if they do not seize the opportunity then some other group, S', will, so that the disruptive re-organization of $N \backslash (S \cup S')$ will take place in any case. Moreover, the reorganization of $N \backslash (S \cup S')$ will take time, and may indeed require the costly creation of new links and so does not occur until a later period. Thus, the members of S may be better off if they seize the opportunity to deviate from $\mathscr{D}$ and profit from the deviation while they can. Thus, for the purpose of analyzing the cooperative stability of a coalition structure in a multi-period game such as that of Section 9.3, the naïve successor function seems the appropriate choice. This corresponds to the idea that, in a large competitive economy, there will be many entrepreneurs who are free to enter by forming new coalitions to exploit available opportunities.

All in all, it seems that the traditional theory of monopoly is misleading, placing far too much stress on its implications for misallocation of resources and too little on its implications for bargaining power and so on the distribution of real income.

9.6. Separations

In Section 9.3, we were concerned with some properties of a stable coalition, with the simplifying assumption that separations occur according to a radioactive-decay rule, which would be appropriate to the extent that they result from demographic change. A few words will be said about endogenous, intended separations. In the process, we recover and extend a key condition from neoclassical economics, that plays a key role in the search-and-matching literature on labor markets: for a stable coalition for production and sale, the wage cannot exceed the value of the marginal product of labor.

In deriving the marginal productivity theory of wages, we recall, John Bates Clark (1899) made a number of assumptions. Some are well known: he assumed perfectly competitive markets and a well-defined, differentiable production function. It is less well known that he assumed that, over some considerable range, employees are indifferent substitutes one for another.[3] We will find that none of those assumptions is necessary; but what is necessary is that the cost of dismissal is zero. For simplicity and generality we return to the expression of the game in TU coalition function form. Thus, the value generated by coalition C is denoted as $v(C)$.

Consider a coalition $C = \{1, 2, \ldots, m\}$ and the deviation $C'_m = \{1, 2, \ldots, m - i\}$. This deviation is equivalent to the dismissal of m. Then the rationality constraint for C'_m is

$$\sum_{i=1}^{m-1} x_i \geq v(C'_m) \tag{9.23a}$$

where, as usual, x_i is the imputation to Agent a_i. If this rationality constraint cannot be met, then coalition C is not stable in the sense of the core of the game, and coalition C'_m will be formed in its place — that is, Agent m will be dismissed. Now,

$$\sum_{i=1}^{m} x_i = v(C) \tag{9.23b}$$

so we have, as a condition for the stability of C,

$$x_m = \sum_{i=1}^{m} x_i - \sum_{i=1}^{m-1} x_i \leq v(C'_m) - v(C) \tag{9.23c}$$

and $v(C'_m) - v(C)$ is the marginal value contribution of Agent m. Thus we have the marginal value contribution as the upper limit of the wage paid to a particular employee (or indeed of the dividend paid to a particular shareholder, etc.). And this does not depend on a market equilibrium or that employees are indifferent substitutes nor indeed, even on the "capitalist" profit-maximizing decision-making of the coalition, but only on the

---

[3]This is often missed in untutored discussions of wages. In these non-specialist discussions, it is fairly commonly supposed that the decreasing marginal product of labor results from individual differences among the employees, so that the "marginal employee" is the one whose skills are least valued. But if there are such individual differences, then the law of one price cannot be applied to labor, and so a marginal productivity principle cannot be applied to determine a price that does not exist! Instead, for Clark, the marginal productivity of labor depends strictly on the number employed, and not at all on any individual differences among the employees. But the marginal productivity of a particular employee may depend on her individual idiosyncracies as well as the number employed, and indeed there is no "law of one price" for labor or for any product in a search market, as Stigler (1961) already noted. Moreover, the importance of individual idiosyncracy is central to the idea of "a good match" in the employment matching literature.

assumption that the dismissal reflects the interests of those who remain in the smaller coalition, whatever their influence on the decision (their bargaining power) may be.

This applies if there are no costs of dismissal. However, costs of dismissal have played an important part in recent discussions of labor markets. Suppose that there is a cost of dismissal, $\psi > 0$, which is charged against the value of C'_m. Then the rationality constraint for C'_m is

$$\sum_{i=1}^{m-1} x_i \geq v(C'_m) - \psi \qquad (9.23\text{d})$$

and in place of (9.23c) we have

$$x_m \leq v(C'_m) - v(C) + \psi. \qquad (9.23\text{e})$$

Thus, the introduction of a cost of dismissal results in an increase in the upper limit of individual wages, and if employees have any bargaining power whatever, this would seem to imply that the realized wage will rise.

9.7.   Chapter Conclusion

An enterprise economy is somewhat more complex than the models typical of cooperative game theory. Nevertheless, an enterprise is a cooperative coalition, in the sense that it is aimed at the mutual benefit of the employees, customers and proprietors. Intertemporal dependencies are important in the economics of the enterprise, but have not been much studied in cooperative game theory. In the previous chapter, some concepts for intertemporal cooperative games were sketched. Enterprises also have members with differentiated roles, and as a result of both this fact and the intertemporal structure of enterprises, a single agent may be a member of more than one coalition. The previous chapter and this one have suggested a definition of presolutions for games in which coalitions may have overlapping memberships: *considerable* imputations and coalition structures.

While an enterprise economy cannot be modeled directly by partitions of the population, there will be limits on the feasible condition structures. Thus, we begin with considerable presolutions to feasible condition structure (FCM) games. For the purposes of a modern theory of enterprise, the limits on the feasible condition structures may be supposed to arise from costly search and matching processes that determine the links among agents. Since search and matching are necessarily noncooperative, we think of each period in the life of the enterprise as comprising a Biform Game in which the search and matching are the noncooperative first stage, and production and sales then take place on a cooperative basis. Modeling the second stage as a BP game, we obtain a conception of the enterprise that contrasts importantly with the neoclassical model of the firm. While monopolistic competition would be the usual case in a model with costly search and matching, no "monopoly restriction" of output is to be expected. Rather, the function of the enterprise will be efficient as among the members of the enterprise, while the distribution of the benefits, allowing for price discrimination and all-or-nothing offers, reflect bargaining power and the rationality constraints that reflect the agents' competitive alternatives.

References

Abbott, B. and L. Costello (1938), Who's on First? (Baseball Almanac). Available at: http://www.baseball-almanac.com/humor4.shtml. Accessed 9 July 2012.

Aivazian, V. A. and J. L. Callen (1981), The Coase theorem and the empty core, *Journal of Law and Economics*, **24**(1), 175–181.

Albizuri, M. J., J. Arin and J. Rubio (2005), An axiom system for a value for games in partition function form, *International Game Theory Review*, **7**(1), 63–72.

Algaba, E., J. M. Bilbao and J. J. Lopez (2001), A unified approach to restricted games, *Theory and Decision*, **50**(4), 333–345.

Aoki, M. (1980), A model of the firm as a stockholder-employee cooperative game, *American Economic Review*, **70**(4), 600–610.

Aumann, R. J. (1985), On the non-transferable utility value, *Econometrica*, **53**, 667–677.

Aumann, R. J. and J. H. Dreze (1974), Cooperative games with coalition structure, *International Journal of Game Theory*, **3**, 217–237.

Aumann, R. J. and M. Kurz (1977), Power and taxes, *Econometrica*, **45**(5), 1137–1161.

Barro, R. J. (1977), Long-term contracting, sticky prices, and monetary policy, *Journal of Monetary Economics*, **3**, 305–316.

Bellman, R. E. (1957), *Dynamic Programming* (Princeton University Press, Princeton, NJ.).

Bishop, R. L (1975), Game-theoretic analyses of bargaining, in *Bargaining* (Ed.) O. Young (University of Illinois Press, Urbana-Champaign), pp. 85–128.

Bolger, E. M. (1989), A set of axioms for a value for partition function games, *International Journal of Game Theory*, **18**, 33–74.

Brandenburger, A. and H. Stuart (2007), Biform games, *Management Science*, **53**(4), 537–549.

Brandenburger, A. M. and H. W. Stuart, Jr. (1996), Value-based business strategy, *Journal of Economics and Management Strategy*, **5**(1), 5–24.

Burdett, K and D. Mortensen (1998), Wage differentials, employer size, and unemployment, *International Economic Review*, **39**(2), 257–273.

Carraro, C. (2003), *The Endogenous Formation of Economic Coalitions* (Edward Elgar, Cheltenham, UK).

Chatain, O. and P. Zemsky (2007), The horizontal scope of the firm: Organizational tradeoffs vs. buyer-supplier relationships, *Management Science*, **53**(4), 550–565.

Clark, J. B. (1899), *The Distribution of Wealth* (Macmillan, New York).

Coase, R. (1960), The problem of social cost, *Journal of Law and Economics*, **3**(1), 1–44.

Coddington, A. (1975), A theory of the bargaining process: Comment, in *Bargaining*, (Ed.) O. Young (University of Illinois Press, Urbana), pp. 219–227.

Cross, J. G. (1975), A theory of the bargaining process, in *Bargaining*, (Ed.) O. Young (University of Illinois Press, Urbana), pp. 191–218.

Debreu, G and H. E. Scarf (1963), A limit theorem on the core of an economy, *International Economic Review*, **4**(3), 235–246.

de Clippel, G. and R. Serrano (2008a), Bargaining, coalitions and externalities: A comment on Maskin, Brown University, Department of Economics Working Paper No. 2008-16.

de Clippel, G. and R. Serrano (2008b), Marginal contributions and externalities in the value, *Econometrica*, **76**(6), 1413–1436.

Diamond, P. A. (1982), Aggregate demand management and search equilibrium, *Journal of Political Economy*, **90**, 881–894.

Domar, E. (1940), Capital expansion, rate of growth and employment, *Econometrica*, **14**, 137–147.

Fellner, W. (1967), Operational utility: The theoretical background and a measurement, in *Ten Economic Studies in the Tradition of Irving Fisher*, in (Ed.) W. Fellner (John Wiley & Sons Inc, New York), pp. 39–75.

Forgo, F., J. Szep and F. Szidarovszky (1999), *Introduction to the Theory of Games: Concepts, Methods, Applications* (Kluwer, Dordrecht).

Funaki, Y. and T. Yamato (1999), The core of an economy with a common pool resource: A partition function approach, *International Journal of Game Theory*, **28**(2), 157–171.

Galbraith, J. K. (1952), *American Capitalism* (Houghton Mifflin, Boston).

Gillies, D. B. (1953), Some Theorems on n-person Games. Doctoral Dissertation, Princeton University, Princeton, NJ.

Gillies, D. B. (1959), Solutions to general non-zero-sum games, in *Contributions to the Theory of Games*, Vol. IV (Annals of Mathematics Studies, Number 40), (Eds.) A. W. Tucker and R. D. Luce (Princeton University, Princeton), pp. 47–86.

Haeringer, G. (1999), Weighted Myerson value, *International Game Theory Review*, **1**(2), 187–192.

Hall, R. E (2005a), Employment fluctuations with equilibrium wage stickiness, *American Economic Review*, **95**(1), 50–65.

Hall, R. E (2005b), Job loss, job finding, and unemployment in the U.S. economy over the past fifty years, *NBER Macroeconomics Annual*, **20**, 101–137.

Hall, R. E. (2008), General equilibrium with customer relationships: A dynamic analysis of rent-seeking, Hoover Institution and Department of Economics, Stanford University; National Bureau of Economic Research, Working Paper.

Hall, R. E. and P. R. Milgrom (2008), The limited influence of unemployment on the wage bargain, *American Economic Review*, **98**(4), 1653–1674.

Harrod, R. (1939), An essay on dynamic theory, *Economic Journal*, **49**, 14–63.

Harsanyi, J. (1963), A simplified bargaining model for the n-person cooperative game, *International Economic Review*, **4**(2), 194–220.

Hart, S. (2004), A comparison of non-transferable utility values, Theory and Decision Library: Series C: Game Theory, Mathematical Programming and Operations Research (Kluwer Academic Publishers, Boston and Dordrecht).

Herings, P. J. J., G. van der Laan, A. J. J. Talman and Z. Yang (2010), The average tree solution for cooperative games with communication structure, *Games and Economic Behavior*, **68**(2), 626–633.

Hicks, J. R. (1932), *The Theory of Wages* (Macmillan, London).

Kaneko, M. (2005), *Game Theory and Mutual Misunderstanding* (Springer-Verlag, Berlin).

Koczy, L. (2007), A recursive core for partition function form games, *Theory and Decision*, **63**, 41–51.

Lange, O. (1942), The foundations of welfare economics, *Econometrica*, **10**, 215–228.

Leibenstein, H. (1969), Organizational or frictional equilibria, X- efficiency, and the rate of innovation, *Quarterly Journal of Economics*, **83**(4), 600–623.

Lewis, W. A. (1942), Notes on the economics of loyalty, *Economica*, **9**(36), 333–348.

Luce, R. D. and H. Raiffa (1957), *Games and Decisions* (Wiley and Sons, New York).

Maskin, Eric (2004), Bargaining, coalitions and externalities, Plenary Lecture, Second World Congress of the Game Theory Society, Marseille.

McCain, R. A. (1980), A theory of codetermination, *Journal of Economics* (*Zeitschift fur Nationalokonomie*), **40**(12), 65–90.

McCain, R. A. (1987), Scalping: Optimal pricing of performances in sports and the arts, *Journal of Cultural Economics*, **11**(1).

McCain, R. A. (2009), *Game Theory and Public Policy* (Elgar, Cheltenham, U.K.).

McCain, R. A. (2010), *Game Theory: A Nontechnical Introduction to the Analysis of Strategy*, Revised Edition (World Scientific Publishers, Singapore).

McKinsey, J. C. C. (1952), *Introduction to the Theory of Games* (McGraw-Hill, New York).

McQuillin, B. (2009), The extended and generalized Shapley value: Simultaneous consideration of coalitional externalities and coalitional structure, *Journal of Economic Theory*, **144**, 696–721.

Michener, H. A., M. S. Salzer and G. D. Richardson (1989), Extensions of Value solutions in constant-sum non-side payment games, *Journal of Conflict Resolution*, **33**(3), 530–553.

Mill, J. S. (1909, reprint edn. 1987), *Principles of Political Economy*, (A. M. Kelley, New York).

Morgenstern, O. and G. Schwödiauer (1976), Competition and collusion in bilateral markets, *Journal of Economics (Zeitschrift für Nationalökonomie)*, **36**(4), 217–245.

Mortensen, D. T. (1982), The matching process as a noncooperative bargaining game, in *The Economics of Information and Uncertainty*, (Ed.) J. J. McCall (University of Chicago Press, Chicago).

Mortensen, D. T. and Christopher Pissarides (1994), Job creation and job destruction in the theory of unemployment, *Review of Economic Studies*, **61**(3), 397–415.

Myerson, R. B. (1976), Values of games in partition function form, *International Journal of Game Theory*, **6**(1), 23–31.

Myerson, R. B. (1977), Graphs and cooperation in games, *Mathematics of Operations Research*, **2**, 224–249.

Nash, J. (1950), The bargaining problem, *Econometrica*, **18**, 155–162.

Nash, J. (1953), Two-person cooperative games, *Econometrica*, **21**, 128–140.

Ng, Y.-K. (1980), *Welfare Economics: Introduction and Development of Basic Concepts* (Wiley Halsted Press, New York).

Peleg, B. and P. Sudholter (2003), *Introduction to the Theory of Cooperative Games* (Kluwer, Dordrecht).

Pen, J. (1952), A general theory of bargaining, *American Economic Review*, **42**, 24–42.

Pham Do, K. H. and H. Norde (2007), The Shapley value for partition function form games, *International Game Theory Review*, **9**(2), 353–360.

Pigou, A. C. (1920), *Economics of Welfare* (Macmillan, London).

Pintassilgo, P. and M. Lindroos (2008), Coalition formation in straddling stock fisheries: A partition function approach, *International Game Theory Review*, **10**(3), 303–317.

Pissarides, C. (1985), Short-run equilibrium dynamics of unemployment, vacancies, and real wages, *American Economic Review*, **75**, 676–690.

Ray, D. and R. Vohra (1999), A theory of endogenous coalition structures, *Games and Economic Behavior*, **26**, 286–336.

Roth, A. (1979), *Axiomatic Models of Bargaining* (Springer-Verlag, Berlin).

Rubinstein, A. (1982), Perfect equilibrium in a bargaining model, *Econometrica*, **50**(1), 97–109.

Saraydar, E. (1965), Zeuthen's theory of bargaining: A note, *Econometrica*, **33**(4), 802–813.

Scarf, H. E. (1967), The core of an n person game, *Econometrica*, **35**(1), 50–69.

Schelling, T. (1960), *The Strategy of Conflict* (Harvard University Press, Cambridge).

Schmeidler, D. (1969), The nucleolus of a characteristic function game, *SIAM Journal on Applied Mathematics*, **17**(6), 1163–1170.

Selten, R. (1964), Valuation of N-person games, in *Advances in Game Theory* (Annals of Mathematics Studies, Number 52), (Eds.) M. Dresher, L. S. Shapley and A. W. Tucker (Princeton University Press, Princeton), pp. 577–626.

Selten, R. (1975), Reexamination of the perfectness concept for equilibrium points in extensive games, *International Journal of Game Theory*, **4**, 25–55.

Shapley, L. S. (1953), A value for n-person games, in *Contributions to the Theory of Games*, Vol. II (Annals of Mathematics Studies Number 28), (Eds.) H. W. Kuhn and A. W. Tucker (Princeton University Press, Princeton), pp. 305–317.

Shapley, L. (1967), Utility comparison and the theory of games, Working Paper, the RAND Corporation. Available at: http://www.rand.org/pubs/papers/2008/P3582.pdf. Accessed 6 July 2012.

Shapley, L. (1969), Utility comparison and the theory of games, *La Decision: Aggregation et Dynamique des Ordres du Preference* (Editions du Centre National de la Recherche Scientifique), pp. 251–263.

Shimer, R. (2005), The cyclical behavior of equilibrium unemployment and vacancies, *American Economic Review*, **95**(1), 25–49.

Sims, C. A. (2003), Implications of rational inattention, *Journal of Monetary Economics*, **60**(3), 665–690.

Solow, R. (1956), A contribution to the theory of economic growth, *Quarterly Journal of Economics*, **70**, 65–95.

Stigler, G. (1961), The economics of information, *Journal of Political Economy*, **69**, 213–225.

Svejnar, J. (1986), Bargaining power, fear of disagreement, and wage settlements: Theory and evidence from U.S. industry, *Econometrica*, **54**(5), 1055–1078.

Telser, L. (1978), *Economic Theory and the Core* (University of Chicago Press, Chicago).

Thrall, R. M. and W. F. Lucas (1963), N-person games in partition function form, *Naval Research in Logistics Quarterly*, **10**, 281–298.

Tirole, J. (2001), Corporate governance, *Econometrica*, **69**(1), 1–35.

Tsebelis, George (1990), Nested Games: Rational Choice in Comparative Politics (University of California Press, Berkeley, California).

van den Brink, R., G. van der Laan and V. Vasil'ev (2007), Component efficient solutions in line-graph games with applications, *Economic Theory*, **33**(2), 349–364.

von Neumann, J. (1959), On the theory of games of strategy, trans. by Sonya Bargmann *Contributions to the Theory of Games*, Vol. IV (Annals of Mathematics Studies Number 40), in (Eds.) A. W. Tucker and R. D. Luce, pp. 13–42 (Princeton University Press, Princeton).

von Neumann, J. and and O. Morgenstern (2004), *Theory of Games and Economic Behavior*, 60th Anniversary Edn. (Princeton University Press, Princeton).

von Wieser, F. (1889), *Natural Value* (Macmillan and Co, London, 1893) trans. by C. A. Malloch (Ed.) (with an introduction by W. Smart).

Williamson, O. E. (1975), *Markets and Hierarchies* (Free Press, New York).

Winter, E. (2002), The Shapley value, in *Handbook of Game Theory with Economic Applications*, Vol. 3, (Eds.) R. J. Aumann and S. Hart (North-Holland).

Zeuthen, F. (1930), *Problems of Monopoly and Economic Warfare* (Routledge and Kegan Paul, London).

Index